WYOMING STATE GEOLOGICAL SURVEY
Gary B. Glass, State Geologist

Report of Investigations No. 51

STRUCTURAL GEOLOGY OF THE LARAMIE MOUNTAINS, SOUTHEASTERN WYOMING AND NORTHEASTERN COLORADO

by

D.L. Blackstone, Jr.[1]

[1]Professor Emeritus,
Department of Geology and Geophysics, University of Wyoming

Laramie, Wyoming

1996

Table of contents

Illustrations

Page

Figures

Plates (back pocket)

Abstract

The Laramie Mountains are a large, asymmetric, doubly plunging uplift with Archean and Proterozoic crystalline rocks exposed in the core. The mountain mass is compartmentalized by northeast-striking shear zones of which the Cheyenne Belt is the most prominent. Tectonic movement varies in direction, but is generally to the east or northeast. Extensional faulting, post-Miocene in age, is superposed on all older structural elements.

Introduction

Purpose

The purpose of this paper is to place the Laramie Mountains in the regional tectonic framework, and describe some structural details associated with the uplift. Major lineaments within the Precambrian core of the range are defined, and how they segment the mountain mass. A tentative chronology of the evolution of the mountains is presented.

Location

The Laramie Mountains lie in the southeastern quarter of Wyoming and extend into northeastern Colorado. The mountains lie between 40° and 43° North Latitude and 105° and 106°30' West Longitude in Albany, Converse, Laramie, Natrona, and Platte Counties, Wyoming, and Larimer and Weld Counties, Colorado (**Figure 1**).

The mountain mass is approximately 140 miles (225 kilometers, km) long and the width of exposure of the exposed core of Precambrian rocks varies from 45 miles (72 km) in the northern segment to 25 miles (40 km) in the southern segment. The highest points topographically are Laramie Peak, 10,272 feet (3131 meters, m), and the Sherman Mountains, 9053 feet (2759 m) (**Figure 1**).

The geologic map of Wyoming (Love and Christiansen, 1985) and the map of the Precambrian basement (Blackstone, 1993b) clearly outline the location and extent of the mountains.

The geometry of the Laramie Mountains is typical of that known throughout the Rocky Mountain foreland. Laramide deformation created crustal warping with long wavelengths (64 miles or 103 km) across the mountain and low amplitudes of approximately 3.1 miles (5 km). Three cross sections across the Laramie Mountains demonstrate the fold shape (**Figure 2**). Part of the differential vertical uplift measured at the top of the Precambrian basement results from movement on reverse faults on the steep side of the asymmetric uplift. Data for the sections is from Blackstone (1993b).

The mountains are the eastern front of the Rocky Mountain foreland deformation and separate the Laramie and Shirley Basins on the west from the Cheyenne-Denver Basin on the east and the Powder River Basin on the north. The area is drained by the North Platte, South Platte, and Laramie Rivers.

The general topography of the eastern flank of the mountains for 30 miles (48 km) north of the Colorado-Wyoming border is shown by a block diagram (**Figure 3**).

Previous investigations

The Laramie Mountains have been investigated intensively by Federal agencies (i.e. U.S. Geological Survey, U.S. Bureau of Mines, U.S. Forest Service), the Wyoming State Geological Survey, and by academicians.

Mapping by graduate students from several universities provides data along the flanks of the range as well as within the Precambrian core. Exploratory drilling off the eastern side of the mountains provides constraint on structural interpretation.

Reflection seismic profiles provide data on crustal characteristics and behavior, in particular data acquired by the Consortium for Continental Reflection Profiling (COCORP) and by the Department of Geology and Geophysics, University of Wyoming.

Specific investigations are referred to in the text, and a more extensive list of previous investigations appears under **References**.

Acknowledgments

The writer appreciates the support of the Department of Geology and Geophysics of the University of Wyoming for office space, clerical support and library facilities. The Wyoming State Geological Survey provided well data and access to electric log files. The writer in particular appreci-

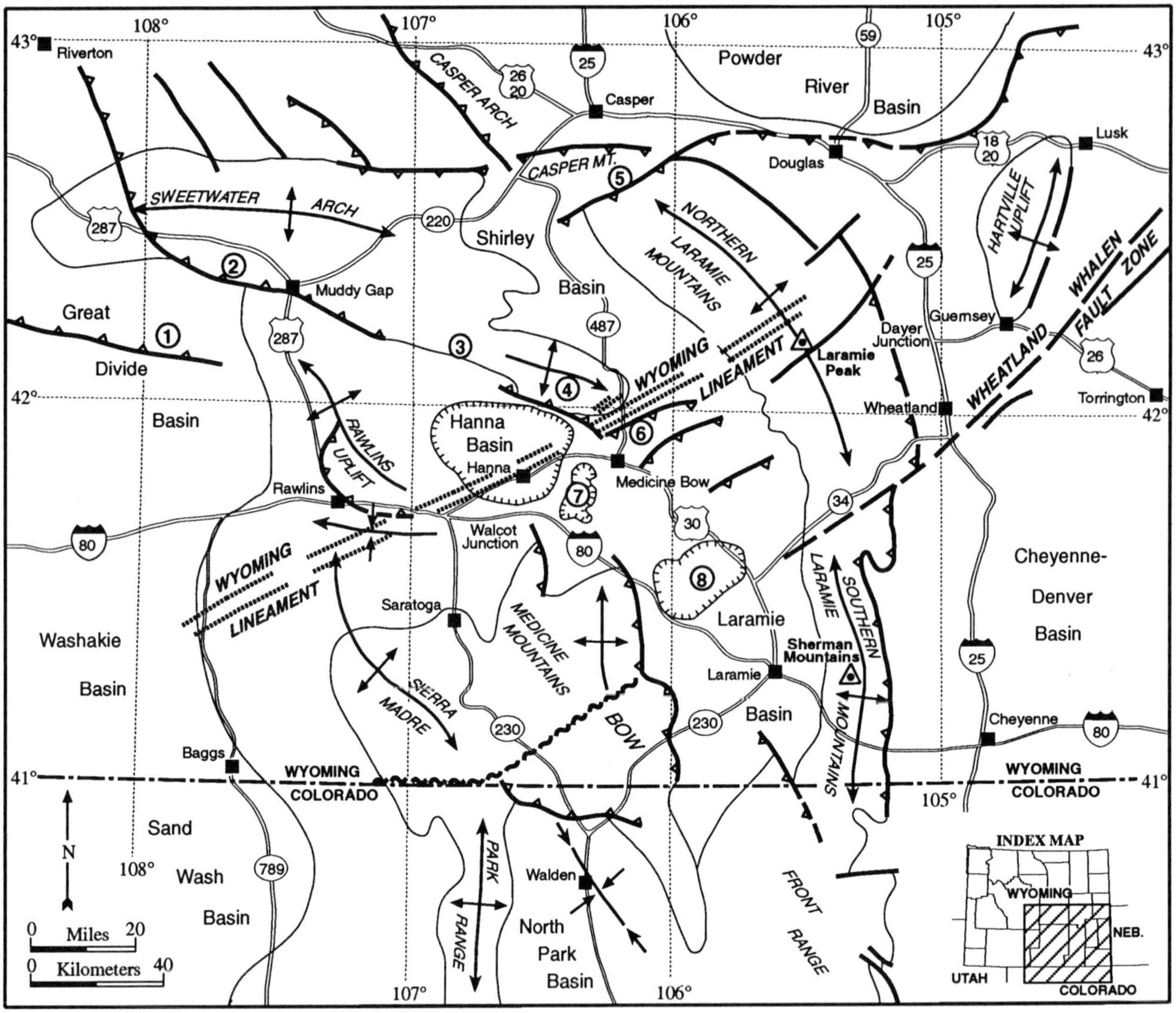

Figure 1. Tectonic index map of southeastern Wyoming and northeastern Colorado. Numbered features are: 1. Wind River thrust fault, 2. Emigrant Trail thrust fault, 3. Seminoe Mountains, 4. Freezeout Hills, 5. Muddy Mountain-Deer Creek fault, 6. Area of northeast-trending folds, 7. Carbon Basin, and 8. Cooper Lake Basin.

ates the cartographic skills of Phyllis Ranz who drafted the illustrations. Discussions with Emmet Evanoff, B. Ronald Frost, Jason Lillegraven, Brainerd Mears, Jr., John Parker, Scott Smithson, and Robert Weimer clarified many points. The writer thanks Arthur Snoke for a critical review of an early draft of the manuscript.

Regional tectonic framework

The major tectonic features of the region are shown on **Figure 1**. The northern end of the Laramie Mountains in part bounds the Powder River Basin and in part lies adjacent to the Hartville uplift. The western flank is bounded by the northeast trending folds associated with the Wyoming lineament and in part by the relatively shallow Laramie-Cooper Lake Basin. The eastern flank of the mountains south of the Wheatland-Whalen fault system borders the Cheyenne-Denver Basin.

The Wyoming lineament (Ransome, 1915) is a zone across which structural style changes character. Northwest of the lineament structures chiefly trend northwest and verge to the southwest. South

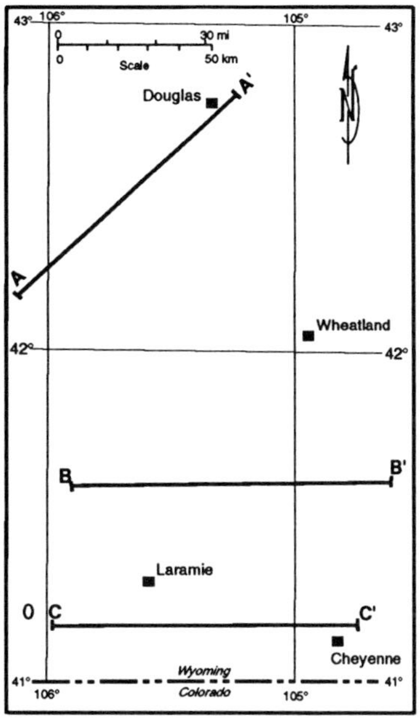

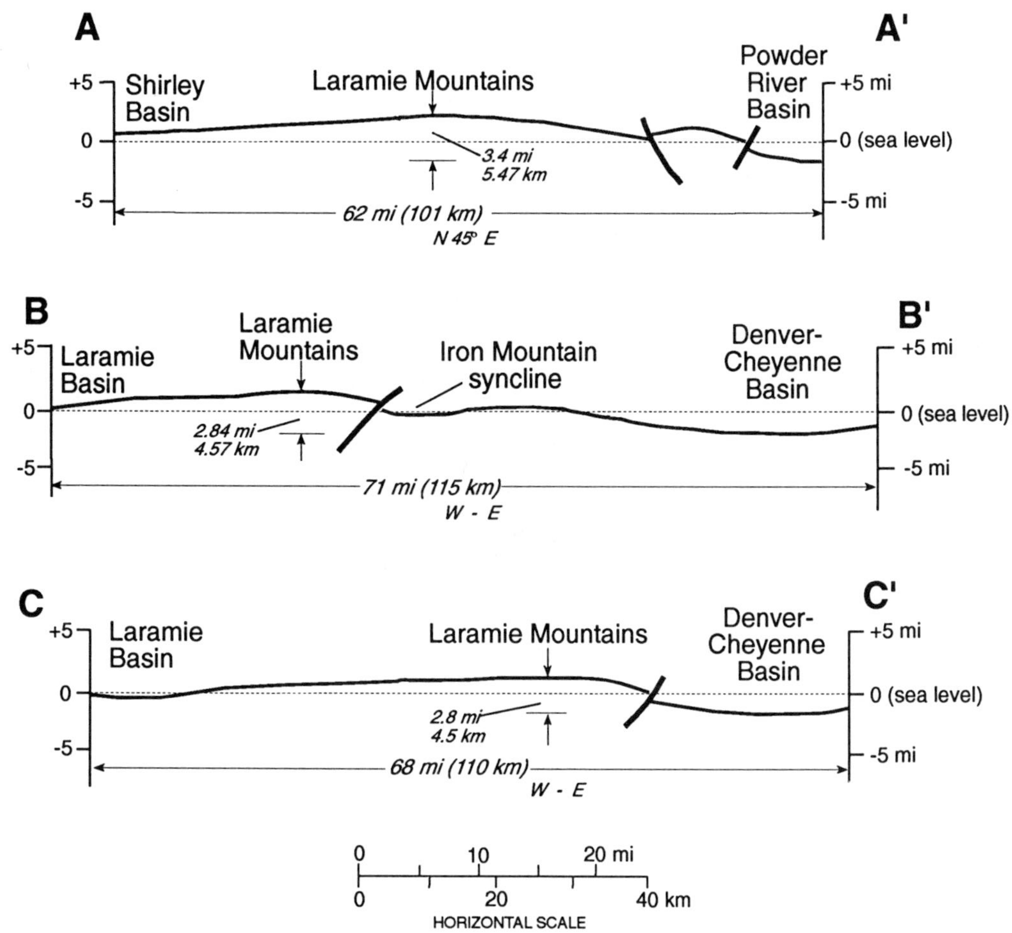

Figure 2. Foreland fold dimensions—wavelength and amplitude, southeastern Wyoming. Profile is at the top of the Precambrian basement or the Phanerozoic-Precambrian unconformity. Map shows approximate locations of cross sections A-A', B-B', and C-C'. Mi, miles; km, kilometers.

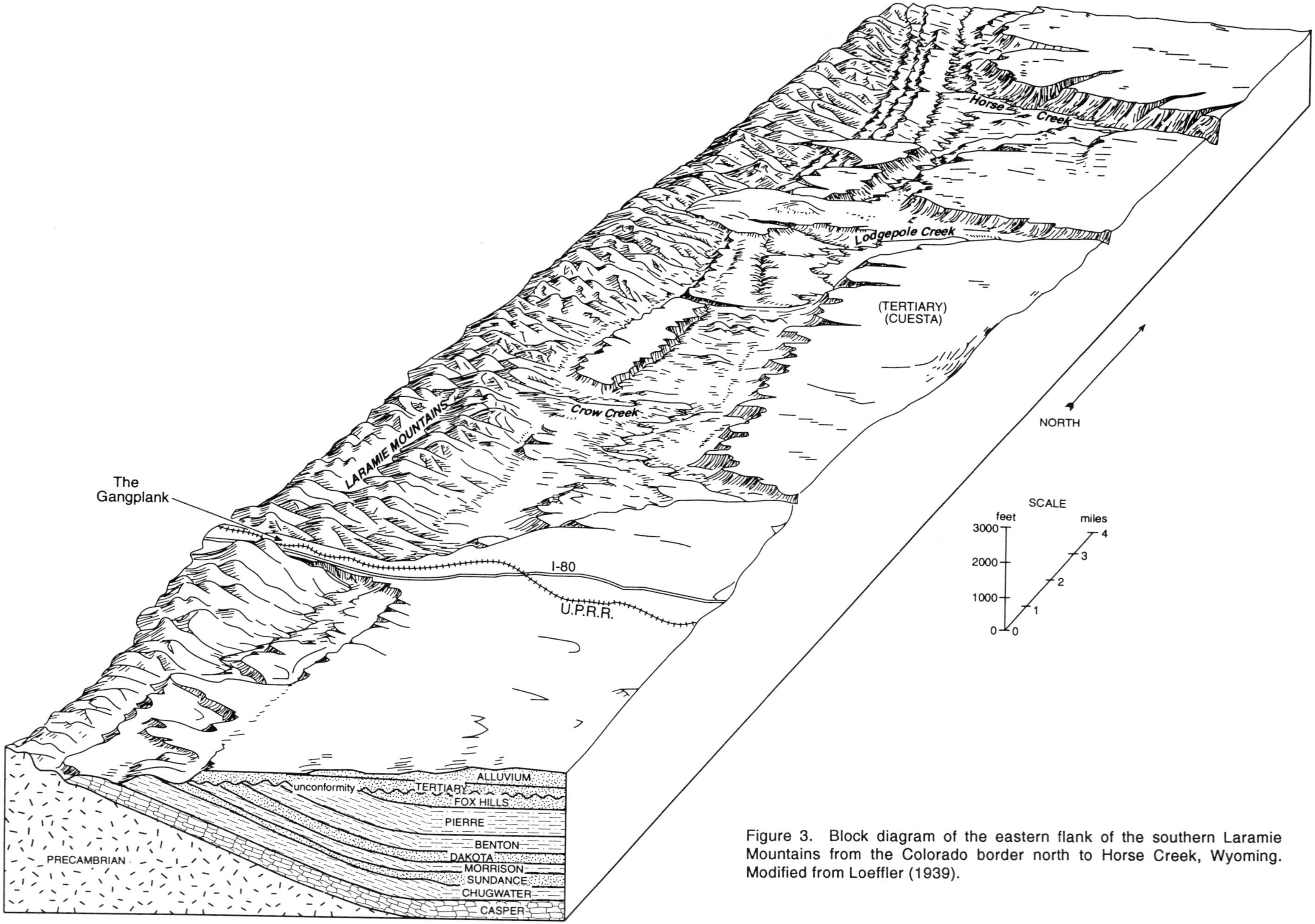

Figure 3. Block diagram of the eastern flank of the southern Laramie Mountains from the Colorado border north to Horse Creek, Wyoming. Modified from Loeffler (1939).

of the lineament structures in general trend north-south and verge to the northwest.

Major northwest-striking zones of structural discontinuity exist in southeastern Wyoming. Geologic mapping in the Medicine Bow Mountains and the Sierra Madre defined a shear zone originally named the Mullen Creek-Nash Fork zone (Houston and others, 1968). Later this major suture was renamed the Cheyenne belt (Karlstrom and Houston, 1984) and was shown to extend from the Sierra Madre, across the Medicine Bow Mountains, across the Laramie Mountains, and as far east as the Hartville uplift. A segment of this shear zone crops out in the Richeau Hills (Mueller, 1982). The suture separates Archean rocks of the Wyoming Province (>2.5 Ga or billion years before present) on the northwest from Proterozoic rocks on the southeast. The suture appears to have been disrupted by the emplacement of the Laramie anorthosite complex (ca. 1.43 Ga).

Intermediate between the Wyoming lineament and the Cheyenne belt area is a series of northwest striking faults associated with the Elmers Rock greenstone belt, defined by Graff and others (1982), and further studied by Snyder (1993) and Chamberlain and others (1993).

The Wheatland-Whalen extensional fault zone is sub-parallel to the Cheyenne belt. The Colorado lineament described by Warner (1978) is a major, broad, northeast striking shear zone along which the Colorado mineral belt occurs. Warner (1978) considered the shear zone to be 75 miles (121 km) wide and to be bounded on the north by what was then described as the Mullen Creek-Nash Fork shear zone.

The Hartville uplift is a structural entity indirectly related to the Laramie Mountains. The uplift trends N35°E (Smith, 1903), is asymmetrical to the southeast, is fault bounded, and is cored by Precambrian basement.

Denson and Botinelly (1949) presented a tectonic map of the uplift showing the configuration of the Precambrian-Paleozoic interface for the area lying northwest of the Wheatland-Whalen fault zone. Drouillard (1963) prepared a tectonic map of the Hartville uplift and the Wheatland-Whalen fault zone, including the intersection of the latter with the Laramie Mountains. Contours were drawn by that writer on top of the Permo-Pennsylvanian Hartville Formation. Drouillard (1963) presented an exaggerated vertical scale cross section (2X) that depicts the Whalen fault as a high angle, southwest dipping reverse fault with a sharp fold in the Paleozoic strata on the toe of the thrust plate. The writer offers an alternate explanation (see discussion of the Grayrocks area, p. 14 below).

The Wheatland-Whalen extensional fault system is approximately 90 miles (145 km) in length and consists of several strands rather than a single fault plane. Fault movement is in general down to the northwest toward the mountains. The normal faults lie in the hanging wall of a thrust fault (the Mule Creek thrust fault to be described later) but no wells penetrate the thrust plate, therefore details of movement on the normal faults are in doubt.

The questions are: (1) are the fault planes listric in form? or (2) are the fault planes planar in form? The normal faults offset rocks as young as middle Miocene Arikaree Formation (McGrew, 1963) but also offset the Precambrian basement. The observed dips of the fault plane range from 63° to 77°, and there are drag dips in Cenozoic strata as large as 30° adjacent to some faults. The writer believes the faults are planar in nature because they cut basement and there is very little reverse dip of Cenozoic strata into the fault planes.

Ahlbrandt and Groen (1987) discussed the structure and evolution of the adjacent Goshen Hole area. They renamed the Wheatland-Whalen fault zone the Sybille lineament. This is misnomer because the fault zone does not cross the Laramie Mountains, nor does it appear along Sybille Creek. They also point out that Chugwater Creek flows in a very straight course (N40°E) from the Laramie Mountains to the town of Chugwater, Wyoming. The stream alignment may be structurally controlled, but available geologic maps show no fault control for the linear stream course (**Plate 1**).

Stratigraphy

The stratigraphic column includes rocks ranging in age from Precambrian (Archean and Proterozoic) to Quaternary. The oldest rocks crop out in the northern segment of the mountains. The basal sandstones exposed at Casper Mountain were long considered to be Cambrian in age and were designated the Flathead Sandstone or Quartzite. Sando and Sandberg (1987) reviewed the evidence and dated the basal sandstones as Middle Devonian, and renamed the strata Fremont Canyon Sandstone. At the southern end of mountains Pennsylvanian age strata unconformably overlie the basement. There is a marked difference in the nature of Cretaceous age rocks on the eastern and western sides of the uplift. **Figure 4** is a graphic presentation of the rocks present in the southern part of the area. **Figure 5** presents columns of the Upper Cretaceous and Paleocene rocks on opposite sides of the mountains. Much of the area on the eastern flank of the range is covered by Cenozoic age rocks unconformably overlying all older rocks, and concealing much of the geologic structure.

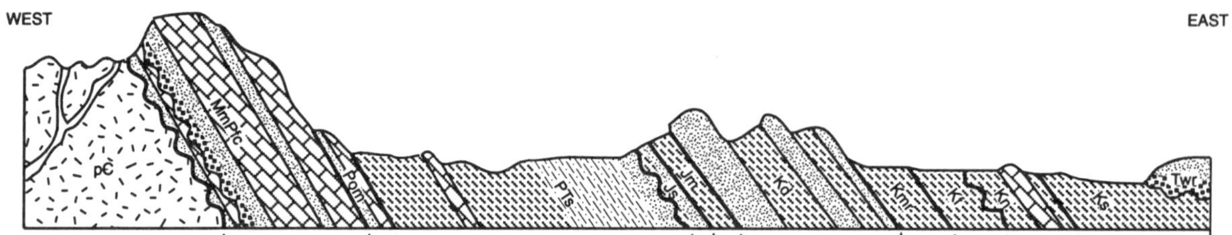

Unit	Sherman Granite	Madison Ls/ Fountain Fm/ Casper Fm	Opeche Sh Minnekahta Ls	Spearfish Fm	Sundance Fm	Morrison Fm	Dakota Group	Mowry Shale	Frontier Fm	Niobrara Fm	Steele Shale	White River Fm
Age	Precambrian	Mississippian/ Pennsylvanian/ Pennsylvanian-Permian	Permian	Permian/ Triassic	Jurassic	Jurassic	Cretaceous	Cretaceous	Cretaceous	Cretaceous	Cretaceous	Oligocene
Thickness (feet)	—	963	32	981	67	232	340	470	355	477	385+	200+
Quality of Exposure	Good	Good	Excellent	Fair	Fair	Fair	Good	Poor	Poor	Fair	Poor	Good

Figure 4. Lithology, thickness, surface expression, quality of exposure, and age of the rocks present in the southern Laramie Mountains, southeastern Wyoming. From Copland (1984).

General structural character

The Laramie Mountains are a large, faulted, eastward verging double-plunging asymmetrical foreland uplift. The mountains are segmented by several northeast striking shear zones including the Garrett shear zone (Langstaff, 1988; Snyder, 1993); the Laramie Peak shear zone (Chamberlain and others, 1993); the Cheyenne belt (**Plate 1**) (Karlstrom and Houston, 1984); and the Wheatland-Whalen fault system (McGrew, 1967). The latter sharply separates the northern and southern segments of the mountain mass (**Plate 1**). The southern segment of the mountains extends southward as the Colorado Front Range (Boos and Boos, 1957).

The exposed Precambrian rocks in the core of the Laramie Mountains consist of crystalline rocks of Archean and Proterozoic age. Rocks lying north of the Cheyenne belt (Karlstrom and Houston, 1984) are of Archean age and were divided by Condie (1969) into three districts: (1) a northern metamorphic complex; (2) the so-called Laramie batholith; and (3) a central metamorphic complex. The age of the rocks in this region is >2.5 Ga (Hills and Armstrong, 1974; Johnson and Hills, 1976). Chamberlain and others (1993) discuss the tectonic history of the Precambrian core of the area and describe two major northeast-striking shear zones: (1) the Laramie Peak zone consisting of two segments: (a) the Garrett-Fletcher Park shear zone and (b) the Cottonwood Park shear zone; and (2) the Cheyenne belt.

The rocks lying south of the Cheyenne belt (Karlstrom and Houston, 1984) are Proterozoic in age, approximately 1.4 Ga. The principal rock masses are the Sherman granite and the Laramie anorthosite complex.

The east flank of the Laramie Mountains is bounded by a series of west dipping thrust faults and

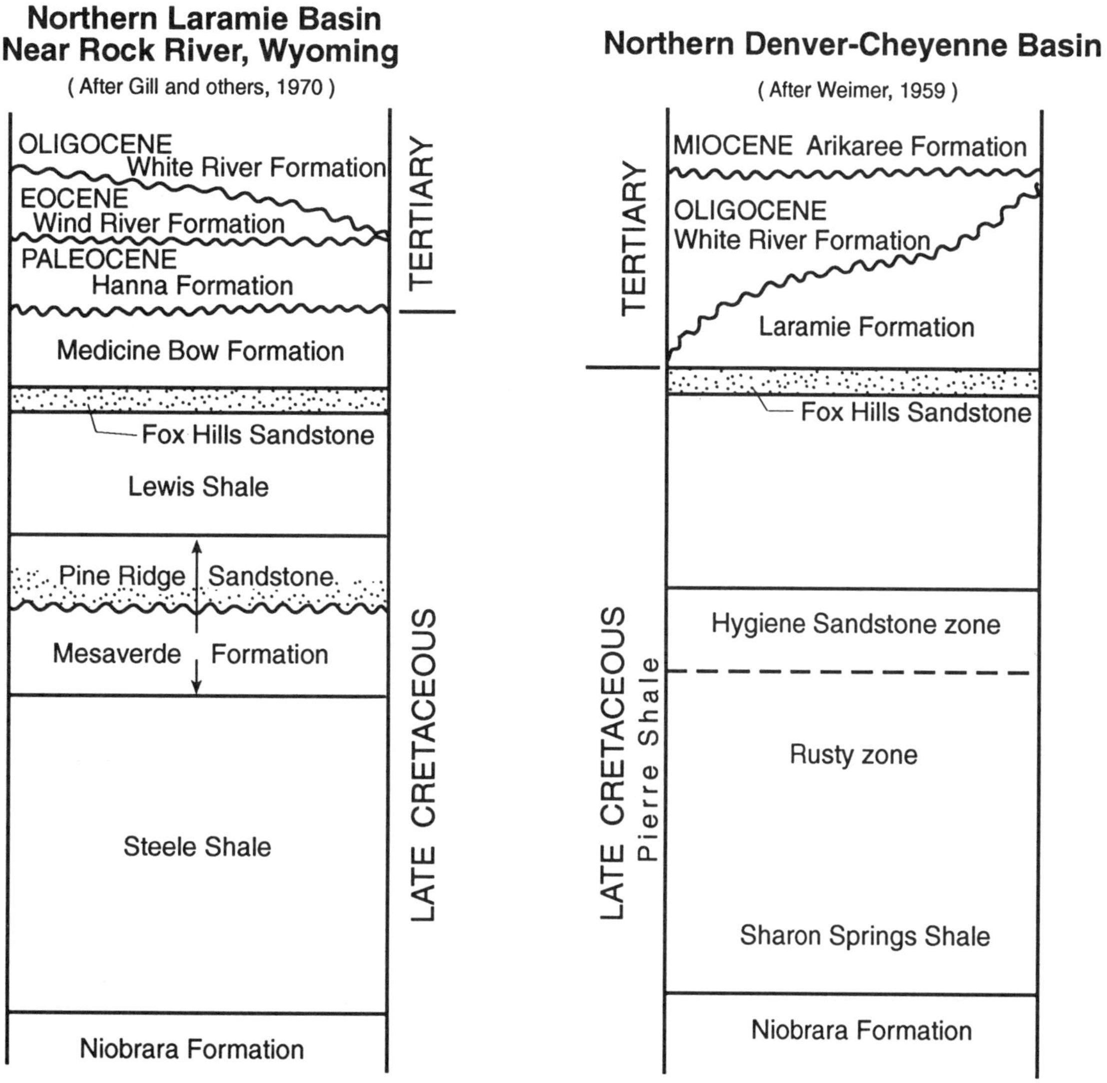

Figure 5. Stratigraphic sections of Late Cretaceous, early Tertiary, and Cenozoic deposits on the west and east sides of the southern Laramie Mountains. Undulating line represents an unconformity; stippled pattern indicates sandstone.

asymmetric folds arranged en echelon, parts of a single system (**Plate 1**).

The names of the folds and faults from north to south are Round Butte (Colorado); Granite Canyon; Mesa Mountain and Crow Creek window; Horse Creek; Iron Mountain fault and syncline; Sheep Mountain fault; and Deadman syncline.

The north-south oriented fault and fold system is intersected in T21N, R70W, by the Wheatland-Whalen fault system. A major northeast striking fault system —the Mule Creek thrust fault— begins here and extends to the northeast at least as far as the Grayrocks area (**Plate 1**) and possibly beyond.

Structural analysis of the southern Laramie Mountains

Details of the geologic structure are presented by means of a series of cross sections oriented transverse to the trend of the mountains, and constructed through critical areas (**Plate 1**).

Sand Creek-Round Butte, Colorado

Cross section I-I' (**Figure 6**) is in northern Colorado in Ts11 and 12N, Rs69 and 70W.

Basic mapping in the area is by Thompson (1938), McGookey (1952), Hunter (1955), Boos and Boos (1957), and Shorey and Howell (1963). At the west end of the cross section a series of north-south trending folds named Sand Creek, Box Elder and Table Mountain involve rocks as young as the Cretaceous Dakota Sandstone. Farther east in T11N, R69W, a well defined anticline, Round Butte, is expressed at the surface in the Cretaceous Pierre Shale.

A deep test, Texaco, Inc. No. 1 Warren Livestock, in section 25, T11N, R69W, drilled to a total depth of 8200 feet (2499 m) and encountered a low angle, west dipping thrust fault at a depth of 5900 feet (1798 m). The surface expression of this fault is concealed by Cenozoic strata. The fault lies some 6 miles (9.7 km) east of the mountain front and does not control the shape of the Laramie Mountain uplift.

Granite Canyon

Cross section H-H' (**Figure 7**) passes west to east through Granite Canyon, Wyoming and the Borie oil field. (McMinn, 1963; Wyoming Geological Association, 1984). There is no thrust fault at the mountain front, but the sedimentary section dips 20°E and passes into a low anticline two miles from the mountain front.

At Borie field the California Company M. King No. 1, section 13, T13N, R68W, reached a total depth of 11,374 feet (3467 m) in Precambrian basement, without faulting. It is possible that the east flank of the anticline is controlled by a west dipping reverse fault but this is not proven by drill records.

Mesa Mountain

Mesa Mountain, underlain by Permian and Pennsylvanian Casper Formation in Ts15 and 16N, Rs69 and 70W (Detail Map 1, **Figure 8**), lies in the hanging wall of the Mesa Mountain thrust fault (Brady, 1949) (cross sections, **Figure 9**). The Casper Formation is folded into an asymmetrical syncline, with modest west dips on the east flank. Prior to erosion an anticlinal roll-over probably existed immediately to the west.

The Crow Creek window in the thrust plate covers about 40 acres (Brady, 1949), exposing Triassic through Cretaceous strata in the footwall. The Precambrian basement and Casper Formation in the hanging wall are folded into a sharp anticline near the toe of the thrust plate.

Structure in the footwall is constrained by a deep exploratory test well, Texaco, Inc. No. 1 Lorenz, in SE NE section 25, T15N, R70W, drilled to a total depth of 8000 feet (2438 m) into Precambrian basement. The well bore (Cross section B-B, **Figure 9**)

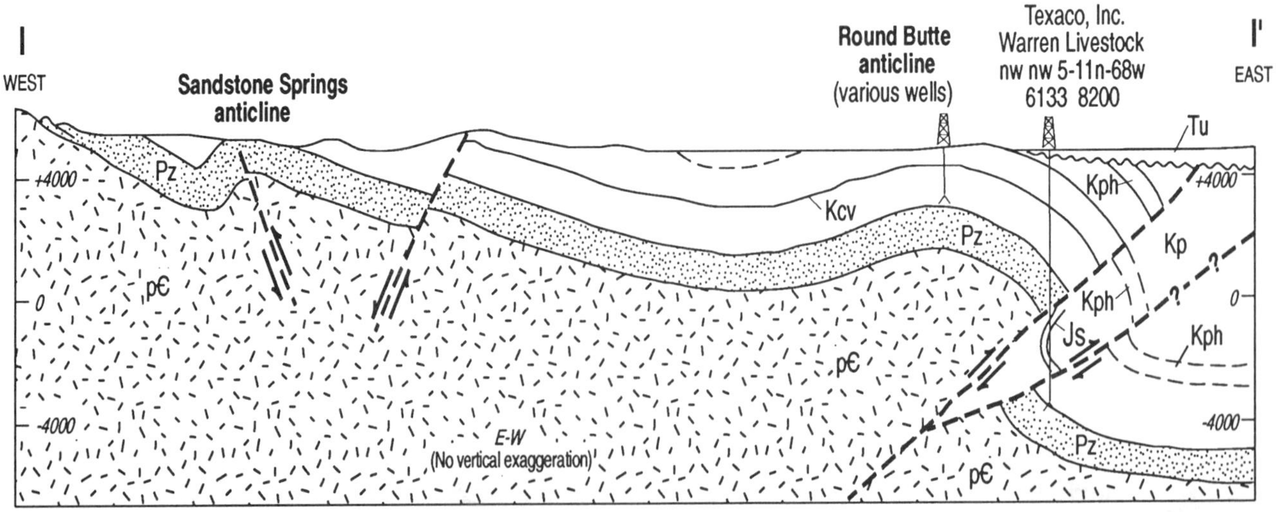

Figure 6. West to east cross section I-I' through Round Butte, Colorado. Location of cross section shown on **Plate 1**. Formation abbreviations are: pЄ, Precambrian basement; Pz, Paleozoic rocks, undifferentiated; Js, Sundance Formation; Kcv, Cloverly Formation; Kph, Hygiene Sandstone Member of the Pierre Shale; Kp, Pierre Shale; and Tu, Tertiary rocks, undifferentiated.

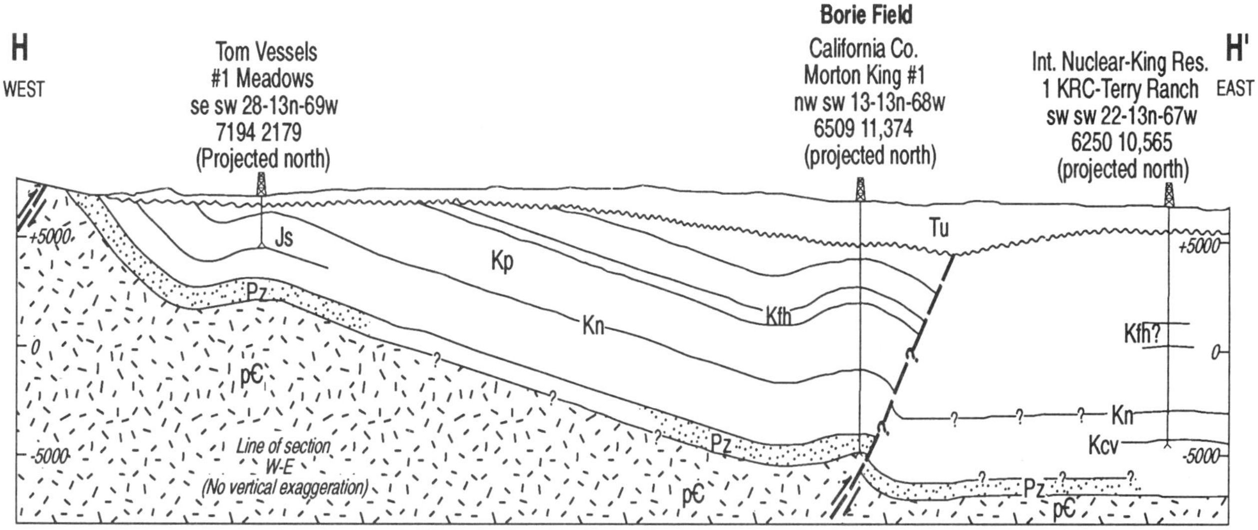

Figure 7. West to east cross section H-H' through Granite Canyon, Wyoming and Borie oil field. Location of cross section shown on **Plate 1**. Formation abbreviations are: p€, Precambrian basement; Pz, Paleozoic rocks, undifferentiated; Js, Sundance Formation; Kn, Niobrara Formation; Kp, Pierre Shale; Kfh, Fox Hills Sandstone; and Tu, Tertiary rocks, undifferentiated.

passed through Precambrian basement, cut the Mesa Mountain thrust fault, drilled in vertical Morrison Formation to a depth of 4000 feet (1219 m) and then crossed a second thrust fault below which a normal section from the Cretaceous Pierre Shale to the Precambrian basement was drilled. The lower fault does not reach the surface and the fault trace is concealed by Cenozoic strata.

Horse Creek

Cross section G-G', (**Figure 10**) near Horse Creek siding on the Burlington Northern Railroad extends from the Precambrian basement eastward to the Horse Creek oil field (Biggs and Espach, 1960; Wyoming Geological Association, 1984). The Precambrian rocks adjacent to this section include anorthosite, granite dikes, and the Iron Mountain titaniferous iron complex.

The Phanerozoic rocks range in age from the Devonian-Mississippian Guernsey Formation to the Cretaceous Fox Hills Sandstone. Original mapping in the area was by Darton and others (1910). At a later date Gray (1947) remapped the area on a planemetric base without topography. Reflection seismic profiles across the area were recorded by COCORP and the Department of Geology and Geophysics, University of Wyoming. The geophysical data will be discussed later.

The mountain flank is bounded by a north-south striking, west dipping thrust fault that places Precambrian basement over Permo-Pennsylvanian Casper Formation (**Figure 10**). A sub-thrust fault dips to the west at a lower angle than the major fault and involves Mesozoic strata at the surface. East of the faulted area the sedimentary rocks dip to the east at angles of 5° to 8° away from the mountain front until interrupted by the Horse Creek anticline.

Iron Mountain syncline and anticline

Iron Mountain syncline is a deep reentrant into the east flank of the Laramie Mountains and is named from the mining camp site, Iron Mountain, in section 27, T19N, T71W. The syncline is asymmetric, bounded on the west by a west dipping thrust fault (Hammond, 1949). Precambrian basement in the hanging wall of them thrust fault (Cross section F-F', **Figure 11**) is in contact with overturned Paleozoic sedimentary rocks in the footwall. The bounding fault extends to the northwest as a zone of cataclastic rocks in Ts19-20N, R71W (**Plate 1**).

COCORP reflection seismic profile Line No. 3, (Allmendinger and others, 1982) crosses the syncline and extends eastward across the adjacent Iron Mountain anticline. The anticline is a broad, low dipping, southeast plunging fold with Precambrian basement exposed in the core. The fold is asymmetric, verges to the east and is broken by the Spring Creek reverse fault (Copland, 1984). Two reverse faults of lesser displacement—the Lambert and Diamond (Copland, 1984)—lie east of and subparallel to the Spring Creek fault.

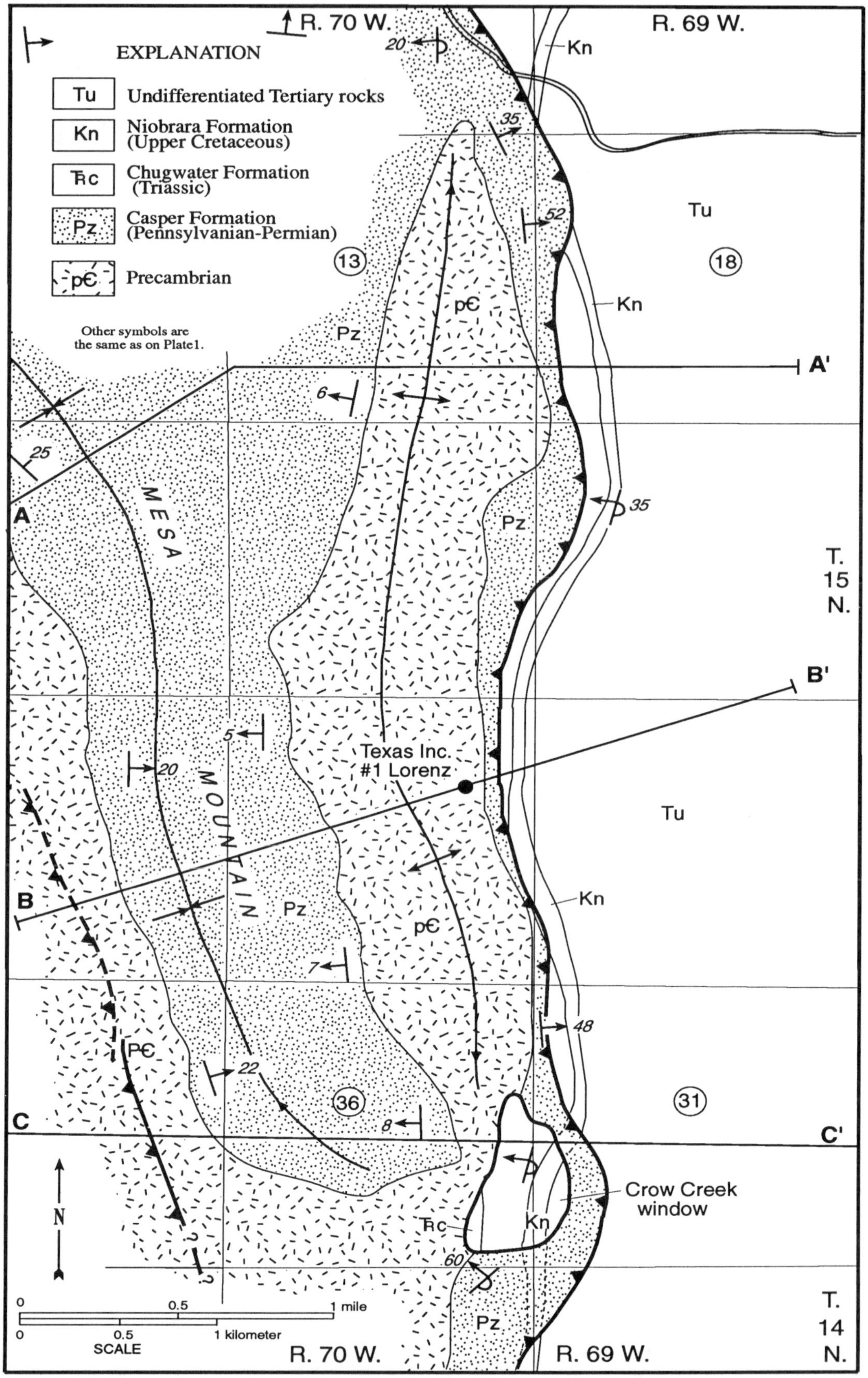

Figure 8. Detail Map 1 of the Mesa Mountain area, showing location of cross sections A-A', B-B', and C-C' (**Figure 9**). Modified from Brady (1949). Symbols are the same as on **Plate 1**.

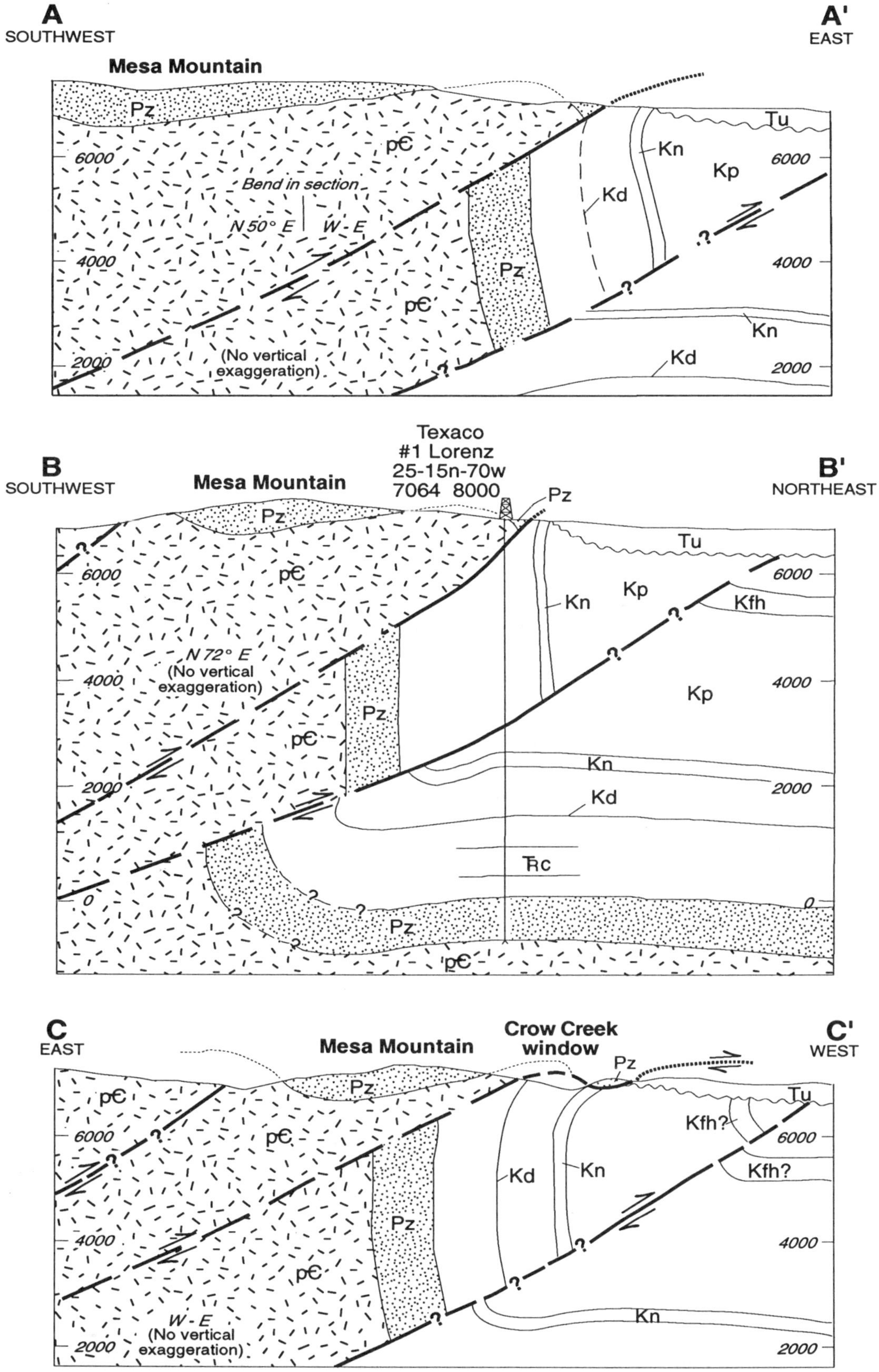

Figure 9. Cross sections A-A', B-B', and C-C' to accompany Detail Map 1 (**Figure 8**). Formation abbreviations are the same as on **Figure 7**, with the addition of: ŦRc, Chugwater Formation; and Kd, "Dakota" sandstone (Cloverly Formation).

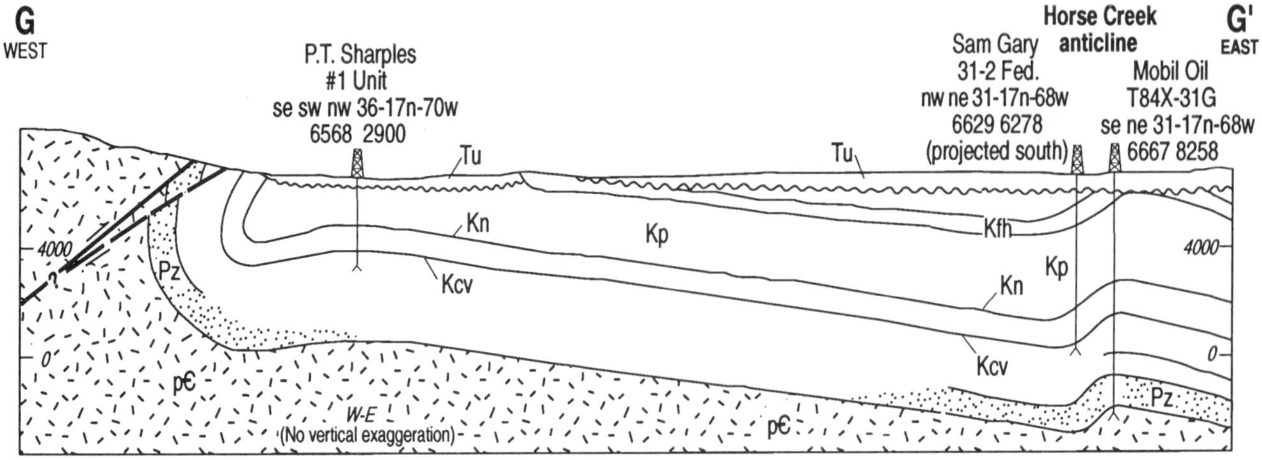

Figure 10. West to east cross section G-G' through Horse Creek siding to Horse Creek oil field. Location of cross section shown on **Plate 1**. Formation abbreviations are: p€, Precambrian basement; Pz, Paleozoic rocks, undifferentiated; Kcv, Cloverly Formation; Kn, Niobrara Formation; Kp, Pierre Shale; Kfh, Fox Hills Sandstone; and Tu, Tertiary rocks, undifferentiated.

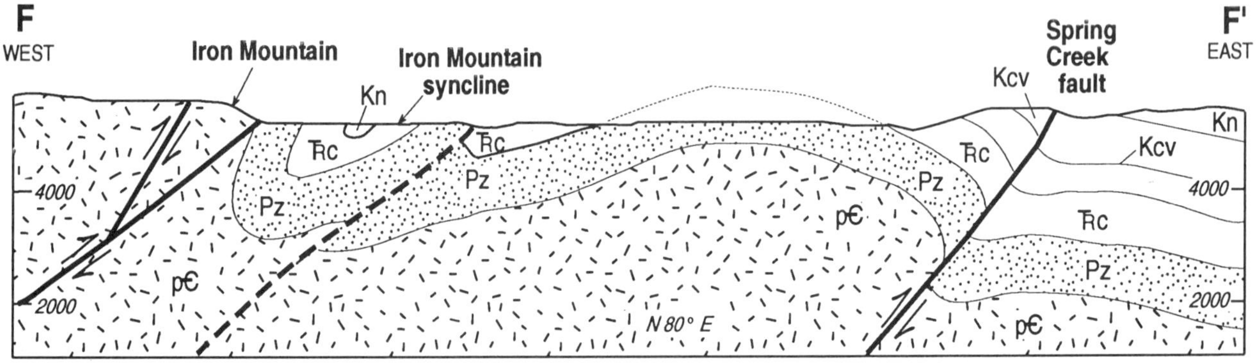

Figure 11. West to east cross section F-F' across Iron Mountain thrust fault, Iron Mountain syncline and anticline, and the Spring Creek fault. Location of cross section shown on **Plate 1**. Formation abbreviations are the same as on **Figure 10**, with the addition of: TRc, Chugwater Formation.

Copland (1984) did not relate the faults that he mapped to those mapped earlier by Haun (1949), McGrew (1967a, b), and Jenkins (1938) to the north.

Sheep Mountain area

Sheep Mountain is at the crest of a high ridge of Permo-Pennsylvanian Casper Formation located in SE section 12, T21N, R70W (McGrew, 1967a). See Cross section E -E', **Figure 12**.

The prominent ridge of Casper Formation lies in the hanging wall of a rather low angle, west dipping, thrust fault, here described as the Sheep Mountain thrust fault. The Precambrian basement and the Pennsylvanian Casper Formation in the hanging wall of the Sheep Mountain thrust fault are in fault contact with the west limb of the broad Sheep Mountain syncline, cored by Permian Goose Egg and Triassic Chugwater Formations. Lateral transport on this fault (Cross section D-D', **Figure 13**) is on the order of 1 mile (0.6 km). The valley of Deadhead Creek roughly parallels the synclinal axis.

Jenkins (1938) mapped and recognized the Sheep Mountain syncline and the anticline to the east but did not recognize thrust faulting. Haun (1949) mapped part of the area and recognized the Sheep Mountain thrust fault. McGrew (1967a) mapped the area as part of an extensive project to define the Wheatland-Whalen fault system and fully defined the northern part of the Sheep Mountain thrust fault, tracing it as far south as section 36, T21N, R70W. A gap in surface mapping exists between work by Hammond (1949), Copland (1984), and McGrew (1967a).

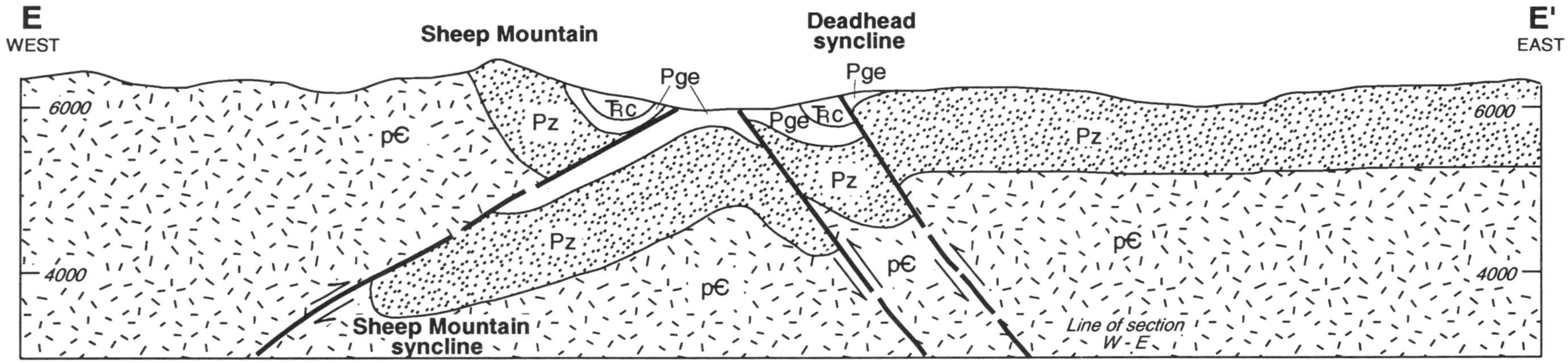

Figure 12. West to east cross section E-E' across Sheep Mountain thrust fault and Deadhead syncline. Location of cross section shown on **Plate 1**. Formation abbreviations are: pЄ, Precambrian basement; Pz, Paleozoic rocks, undifferentiated; Pge, Goose Egg Formation (Permian); and Rc, Chugwater Formation.

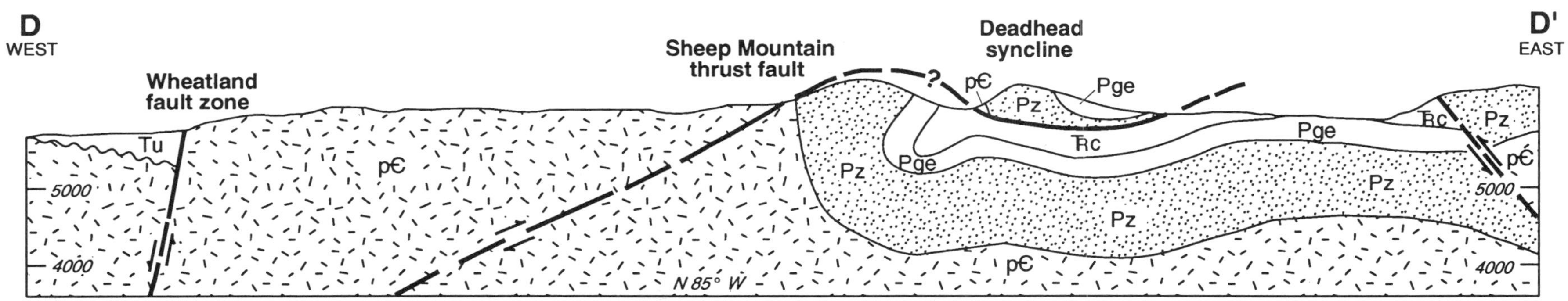

Figure 13. West to east cross section D-D' across the Wheatland fault zone, Sheep Mountain thrust fault, and Deadhead syncline. Location of cross section shown on **Plate 1**. Formation abbreviations are the same as on **Figure 12**.

Deadhead Creek-Mule Creek Junction area

Deadhead Creek is a major stream flowing north across the Natwick Southwest 7 1/2-minute topographic Quadrangle (McGrew, 1967a). The geology was originally presented by Lynn (1947), Haun (1949), and later by McGrew (1967a). The junction of several features occurs near the common corner of Ts21 and 22N, Rs69 and 70W, and they are exposed in the valley and canyon of Deadhead Creek.

The north-south trending Sheep Mountain thrust fault (this paper) intersects the N45°E striking Wheatland -Whalen fault system at an angle of 35°. The latter fault system offsets the Precambrian basement and the unconformably overlying White River and Arikaree Formations. The normal faults are generally down to the northwest toward the mountain uplift.

A low-angle thrust fault here designated the Mule Creek thrust fault appears near the SW corner of section 1, T21N, R70W, and continues to the northeast, striking N45°E (**Plate 1**). Slip on the fault plane is indeterminate because all sedimentary rocks have been stripped from the hanging wall. On the outcrop, Precambrian basement is thrust over the lower part of the Permo-Pennsylvanian Casper (Hartville) Formation and the fault trace is sinuous indicating the low dip of the fault plane. The fault continues to the northeast into the Richeau Hills 7 1/2-minute Quadrangle (McGrew, 1967b) (**Figures 14 and 15**). Along strike the fault trace is cut off by the younger extensional Wheatland normal fault system. The Mule Creek fault plate is deeply dissected resulting in several small klippen of Precambrian basement lying on rocks as young as the Cretaceous Mowry Shale (McGrew, 1967a, b). Mesozoic strata in the footwall are tightly folded with fold axes at an acute angle to the fault strike.

Several areas contain large blocks of basal Cloverly conglomerate (McGrew, 1963). These exotic blocks lie south of the trace of the Mule Creek thrust fault and must have been derived from the hanging wall of that fault. The blocks probably traveled only a short distance and give indirect evidence of the amount of fault slip.

Mule Creek to Grayrocks area

The low dipping thrust fault that crops out in T21N, R70W was designated the Mule Creek thrust earlier in this paper. Mule Creek flows north across the fault zone and the fault is well exposed in that drainage. Lynn (1947) first recognized the fault, and it was later mapped in greater detail by McGrew (1967a, b)

Precambrian basement in the hanging wall of the thrust overides rocks as young as Cretaceous Mowry Shale in the footwall. Cross section A-A' (**Figure 16**) demonstrates the fault relationships in Ts22 and 23N, Rs69 and 70W (see **Plate 1**). Northeast of this area essentially horizontal late Cenozoic strata conceal all the older rocks. Precambrian basement and overlying Paleozoic sedimentary rocks are exposed in a limited area in T25N, R66W (McGrew, 1963).

The writer believes that low angle thrusting exists for a distance of 40 miles (64 km) from the Deadhead-Mule Creek Junction to beyond the Grayrocks area. Structural relief of 6000 feet (1829 m) requires either very sharp folding or crustal shortening and elevation accomplished by thrust faulting.

The problem is further complicated by the Wheatland-Whalen zone of extensional faulting superposed upon the thrust fault effectively cutting off the root of the thrust plate. Only the toe of the hanging wall plate, devoid of any sedimentary cover, is preserved, thereby limiting the critical evaluation of the amount of tectonic transport. The writer estimates that tectonic transport on this thrust fault is approximately 2 miles (3.2 km) to the southeast.

The direction of tectonic transport is approximately S50°E, normal to the strike of the fault plane. There is no hard evidence of extensive oblique or strike slip motion. The motion is at an acute angle (40°) to the direction of motion on the faults bounding the southern segment of the Laramie Mountains. This divergence will be discussed later.

Grayrocks area

The Grayrocks area (McGrew, 1953) takes its name from the abandoned rural Post Office of that name once located in the NE section 14, T25N, R66W, on the Laramie River. The site is shown on the old 1:125,000-scale Hartville topographic quadrangle and on what is now the Register Cliff 7 1/2-minute Quadrangle sheet. The name derived from the gray cliffs of late Tertiary sandstones. The limited outcrops of Precambrian basement and Paleozoic sedimentary rocks in sections 22, 27, and 28,

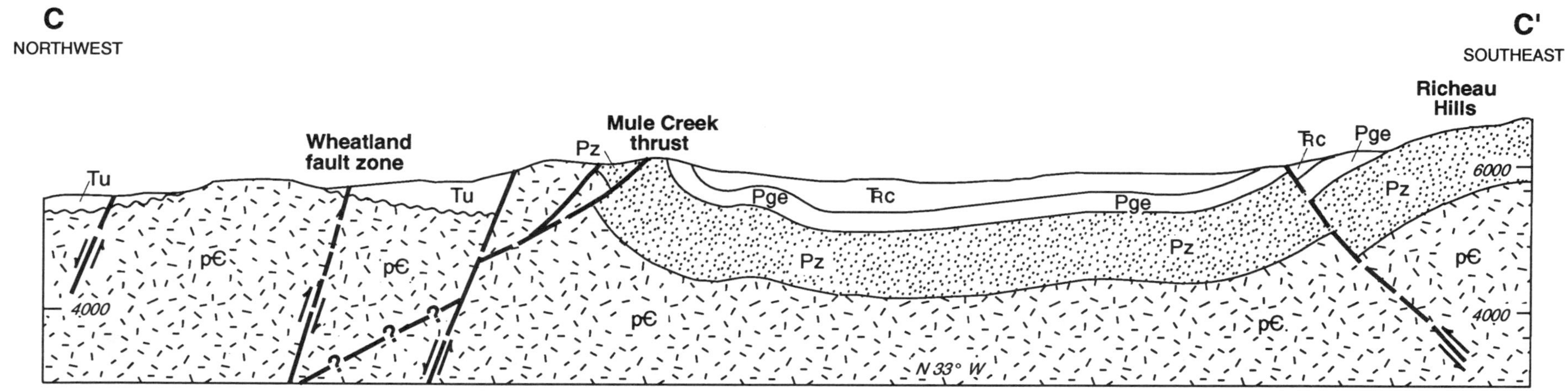

Figure 14. Northwest-southeast cross section C-C' across the Wheatland extensional fault zone and the Mule Creek thrust fault to the Richeau Hills. Location of cross section shown on **Plate 1**. Formations abbreviations are: pЄ, Precambrian basement; **Pz**, Hartville Formation; Pge, Goose Egg Formation; TRc, Chugwater Formation; and Tu, Tertiary rocks, undifferentiated.

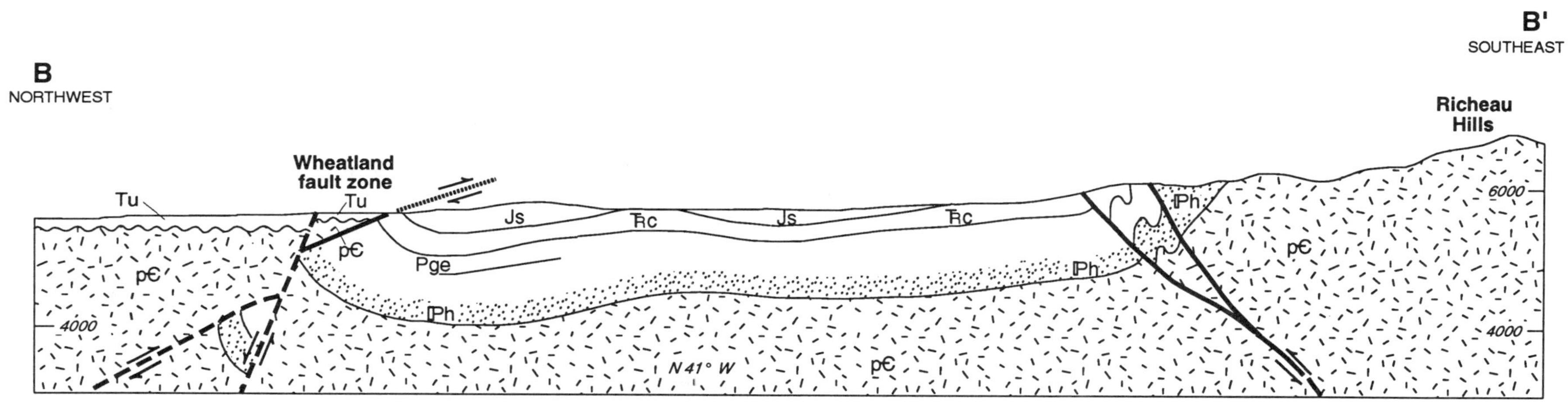

Figure 15. Northwest-southeast cross section B-B' across the Mule Creek thrust to the Richeau Hills. Location of cross section shown on **Plate 1**. Formations abbreviations are the same as on **Figure 14** with the addition of: Js, Sundance Formation.

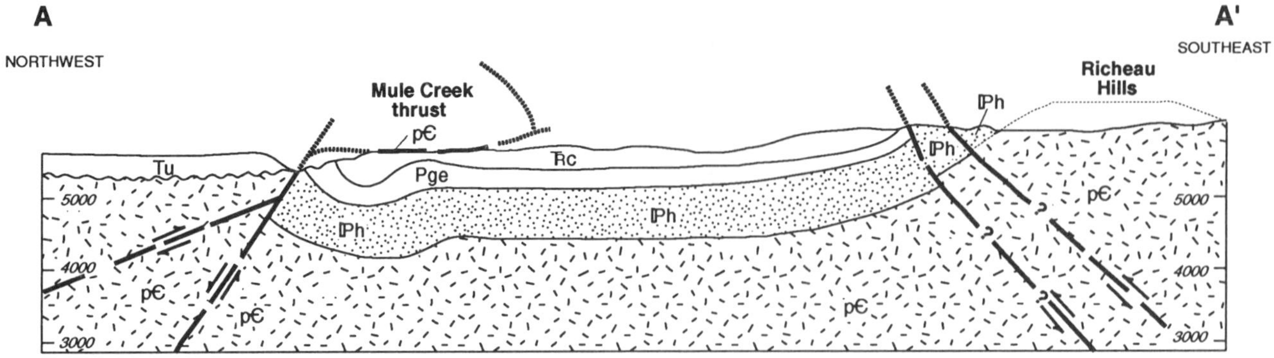

Figure 16. Northwest-southeast cross section A-A', Mule Creek thrust fault to the Richeau Hills. Location of cross section shown on **Plate 1**. Formations abbreviations are the same as on **Figure 14**.

T25N, R66W, lie 2 1/2 miles (4 km) to the southwest of the old Post Office and were described by Smith (1903) as follows:

In addition to these occurrences there are several small areas of schist a little south of the Laramie River about the middle of the quadrangle.

McGrew (1953) mapped the entire Grayrocks area but designated the critical area around the Precambrian outcrops the Moonshine Hills, and the term appears later in McGrew (1963).

The subsurface geology in the Grayrocks area is shown on Detail Map 2 (**Plate 2**), contoured on top of the "Dakota" sandstone (Cloverly Formation). Several wells reach the Precambrian basement to provide control for the basic structure. The major structural element is a broad anticline-syncline pair trending approximately N45°E, parallel to the Wheatland extensional fault zone, but concealed by the unconformable late Cenozoic deposits.

The syncline is asymmetric with a steep northwest limb. At the Moonshine Hills area (Cross section C, **Plate 2**) the Casper Formation dips 45°SE and is overridden by Precambrian basement in the hanging wall of a thrust fault. A similar situation is postulated for Cross sections A, B, and D (**Plate 2**), though there are no surface exposures of the older rocks to verify the assumption.

The thrust fault seen here is the extension of the Mule Creek thrust fault described above. The unusual aspect of this thrust fault is that the tectonic transport is oriented S45°E. Movement on the thrust faults bounding the east flank of the Laramie Mountains, in the region south of the Wheatland fault zone is essentially west to east. The reason for the marked change in transport direction will be discussed later.

At the Moonshine Hills locality, normal faults of the Wheatland fault system are generally down to the northwest and offset the basement and the Paleozoic strata. The amount of movement on these faults can not be definitely established but probably is on the order of 500 to 700 feet (152 to 213 m). McGrew (1963) reports movement on faults in the Wheatland zone to be of the same order of magnitude.

The cross sections (**Plate 2**) show the Mule Creek thrust offset approximately 1000 feet (304 m), which may be in excess of the actual movement.

Structural analysis of the northern Laramie Mountains

Casper Mountain anticline and fault

Casper Mountain, located 5 miles (8 km) south of Casper, Wyoming, is the northerly segment of the Laramie Mountains (**Plate 1**). The uplift is a doubly plunging, asymmetrical, north verging anticline bounded by the south dipping Casper Mountain reverse fault. Precambrian basement is exposed in the core of the fold. Topography, surface geology, and cross sections are presented by Jenkins and Rea (1978) and in a modified form in this report.

The Precambrian rocks in the core of this fold are discussed in detail by Gable and others (1988). The rocks are Archean and Proterozoic age, dated respectively at 2.8 Ga by Peterman and Hildreth (1978) and 1.7 Ga by Hills and Armstrong (1974).

The Casper Mountain fault, a high angle reverse fault (see **Figure 17**) strikes N85°E and dips to the south. Movement on the fault plane places Precambrian basement in fault contact with slightly overturned rocks as young as the Cretaceous Frontier Formation in the footwall.

Muddy Mountain syncline and fault

Casper Mountain anticline is separated from the major part of the northern Laramie Mountains by the Muddy Mountain syncline (**Plate 1** and **Figure 17**). The southern flank of the syncline is defined by a high angle reverse fault in part called the Muddy Mountain fault (Schwarberg, 1959) and farther east called Deer Creek and Little Deer Creek by Berryman (1942) and Little Deer Creek by Sears (1949). The fault system continues to the east for 50 miles (80 km) and increases in displacement (Blackstone, 1988). In the vicinity of Muddy Mountain the fault dips steeply to the southeast, and locally the Precambrian basement is in fault contact with Pennsylvanian Casper Formation.

Northern bounding fault

The primary, most continuous fault in the northern Laramie Mountains is that separating the north-northeast flank of the mountains from the Powder River basin. Segments of the fault have been given separate names (or none) beginning with the Muddy Mountain fault (Schwarberg, 1959); Deer Creek (Berryman, 1942) and Little Deer Creek (Sears 1949); Box Elder (Drwenski, 1952); and South Glenrock (Curry and Curry, 1972). The same fault continues to the east without a name, concealed by the unconformable White River Formation to the vicinity of Douglas, Wyoming. The use of several names for the same fault is confusing, therefore the writer proposes to name this fault the Northern Bounding Fault (NBF).

The western segment of the fault strikes N60°E for 30 miles (48 km) to a point south of Glenrock, Wyoming, then changes strike to N80°W for a distance of 32 miles (51 km) to a point near Orin Junction, Wyoming and is concealed by the Oligocene White River Formation (**Figure 18**). Near Orin Junction the fault has no formal name. Blackstone (1988) discussed this fault and described the occurrence at three localities.

The writer reviewed data obtained from exploratory wells drilled for oil or gas to constrain the location of the fault and from this data prepared a map showing the extent of the fault (**Plates 1 and 3**) and Cross sections A through F, **Plate 3**. Recent drilling near Douglas, Wyoming confirms the magnitude of this fault. The critical well, D.D. Hendrickson 34-19 Scott Federal in SE SE section 19, T32N, R70W, drilled to a total depth of 15,200 feet (4633 m), passed through Precambrian basement, and bottomed in Casper Formation in the footwall (Cross section E, **Plate 3**).

West and East La Prele anticlines

The north and northeastern flanks of the northern Laramie Mountains (**Plate 3**) are characterized by faulted folds such as the West and East LaPrele anticlines (Barlow, 1950), La Bonte, Sheep Mountain, Poison Lake, and Crooked Creek anticlines (Spelman, 1959). These folds trend northwest and are bounded by reverse faults on the southwest flanks. The bounding faults are in the nature of "back thrusts" relative to the NBF. The situation at the West LaPrele anticline (Barlow, 1950) is very similar to that seen at Iron Mountain on the east flank of the southern Laramie Mountains as described earlier.

West La Prele anticline (Cross section B, **Plate 3**) is a large, southeast plunging fold with no western closure, and is bounded on the south flank by a high angle reverse fault. The movement on this fault is small, only about 200 feet (61 m). West La Prele anticline is separated from East La Prele by a northwest plunging syncline and possible associated faulting.

Sheep Mountain anticline

Sheep Mountain is a large, discrete, doubly-plunging anticline with Precambrian basement exposed in the core. The fault strikes N60°W and dips approximately 50°N but the fault plane is not well exposed; therefore, determination of the fault dip is at best an approximation. Fault displacement decreases rapidly in both directions down plunge from the crestal area. The Sheep Mountain bounding fault is concealed to the west by the unconformably overlying Oligocene White River Formation. Spelman (1959) considered the Sheep Mountain

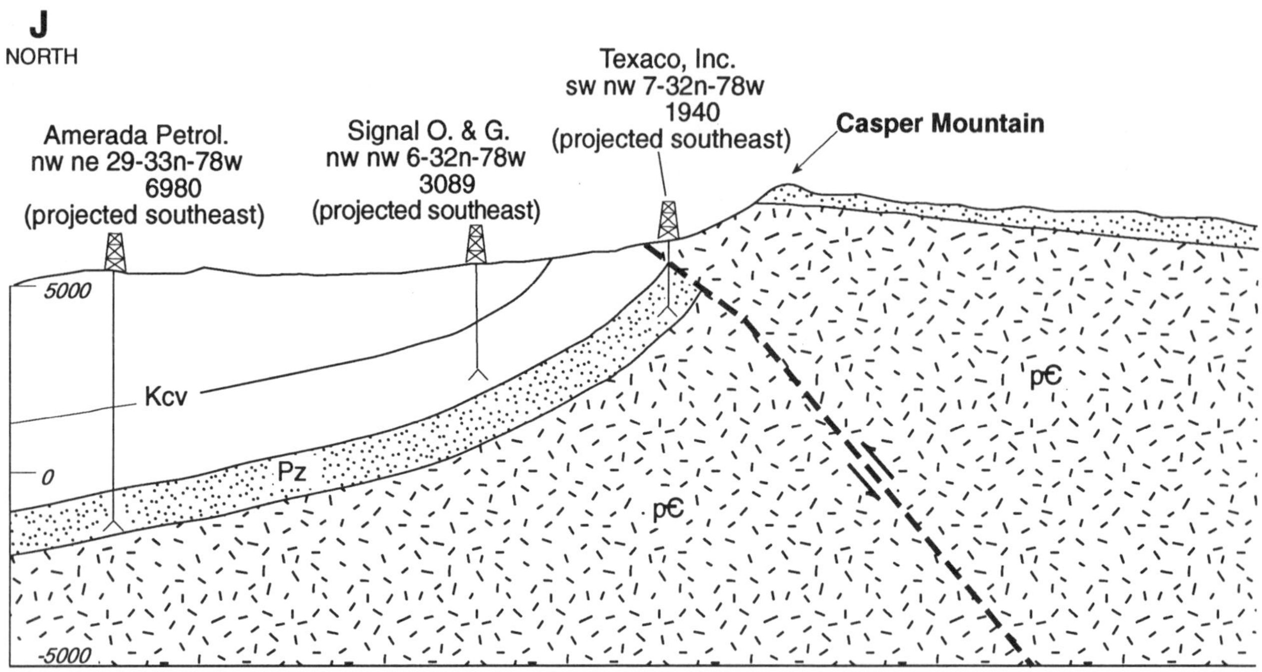

Figure 17. North-south cross section J-J' across Casper Mountain. Location of cross section shown on **Plate 1**. Formation abbreviations are: p€, Precambrian basement; Pz, Casper Formation; and Kcv, Cloverly Formation.

fault to be the eastward extension of the fault bounding the south flank of the West La Prele anticline. The writer does not accept this interpretation and believes that Sheep Mountain anticline lies to the southwest of the East La Prele anticline, and that the controlling fault is separate and distinct from the West La Prele fault (see Detail Map 3 and Cross section C, **Plate 3**).

Associated en echelon folds

Four northwest trending asymmetric anticlines occur in T31N, R71W, in the vicinity of Sheep Mountain. La Bonte anticline lies in T31N, R71W, about 5 miles (8 km) south of Douglas, Wyoming. The fold is asymmetric to the southwest with Triassic Chugwater Formation exposed in the core along the valley of the North Platte River. A warm water spring occurs on the crest of the fold (**Plate 3**).

Crooked Stick anticline, a fold with low structural relief, lies between La Bonte and Sheep Mountain anticlines but is not faulted at the surface.

Another group of folds exists in T30N, R72W, southwest of Sheep Mountain. The northern fold in this group has two names—Poison Lake and Sage Hen. The first name is given preference, The southern fold is Little Sage Hen anticline. Both folds are asymmetric to the southwest and are bounded by north dipping reverse faults. The faults dip to the northeast in a direction opposite to that of the Northern Bounding Fault (NBF). Such faults have been referred to as "back thrusts" and may be controlled by movement on the NBF.

Wheatland-Esterbrook area

The northeastern margin of the Laramie Mountains lies along a line trending N30°W between Esterbrook and the junction of Wyoming State Highway 34 and Interstate Highway 25. (**Figure 1** and **Plate 1**). In this area the contact between Precambrian basement and the overlying Paleozoic sedimentary rocks is concealed by the unconformably overlying late Cenozoic sediments. The questions are: (1) is there a normal sedimentary-Precambrian contact dipping to the east? or (2) is the contact a southwest dipping fault between basement and the overlying rocks?

Near Dwyer Junction (T27N, R68W) several wells penetrated Paleozoic sedimentary rocks that dip to the west at low angles (5° to 10°), suggesting a syncline before the Precambrian core of the range is reached. In view of the fault pattern to the south, the writer suggests that the mountain boundary here is a fault.

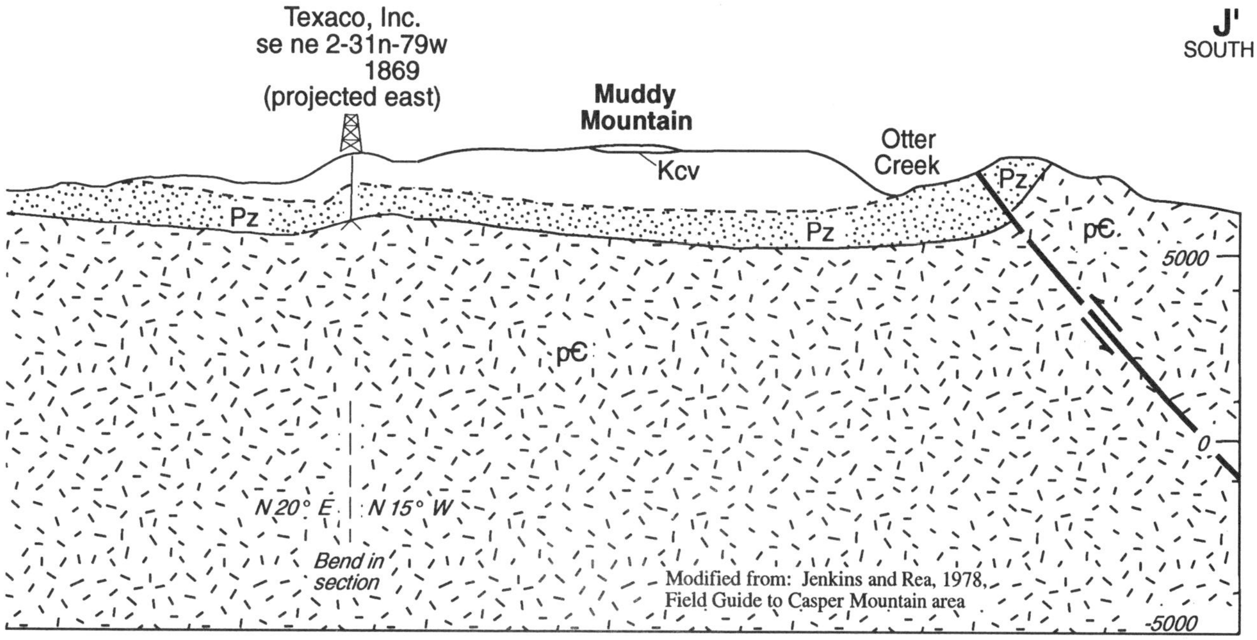

Figure 17. *Continued.*

Garrett-Orin Junction fault

A major shear zone in the core of the northern Laramie Mountains was named the Garrett shear zone by Snyder (1993) who mapped two faults northward to the contact between Precambrian basement and the overlying Paleozoic sediments; he further suggested that the fault might extend farther to the northeast. Drilling in the vicinity of Orin Junction, Wyoming verifies this suggestion. The Orin Junction fault strikes N40°E, appears to dip very steeply to the west, and displacement may be as much as 4000 feet (1219 m), down to the west (Cross section G, **Plate 3**).

The Garrett-Orin Junction fault system is Laramide in age and probably results from reactivation of a shear zone of Precambrian age. The fault lies in the hanging wall of the NBF and has been carried northward by movement on that fault.

Geophysical investigations

Geophysical investigations in the Laramie Mountains proper and on the flanks have been of two types. The first type included gravity measurements designed to define the size, shape, and position of the base of the Laramie anorthosite complex. Gravity measurements were made by Hodge (1973), Iltis (1983), and Johnson and Smithson (1983).

The second type of investigation was designed to determine the deep nature of the Laramide fault system at the mountain margin and the deep crustal structure, by means of reflection seismic methods. The original investigation was part of a COCORP (Consortium for Continental Reflection Profiling) project dealing with foreland structure. Three COCORP reflection seismic profiles were recorded in the Laramie Mountains (**Plate 1**).

The first part of the COCORP investigation involved interpretation of the mountain margins, including thrust faults by reflection profiles along lines 1 through 3 (Brewer and others, 1982). The investigators concluded that the uplift of the southern Laramie Mountains was accomplished by displacement along a set of west dipping thrust faults activated by deep lateral compression.

The second part of the COCORP investigation involved probing the deeper structure within the Precambrian basement and its effect on the Laramide

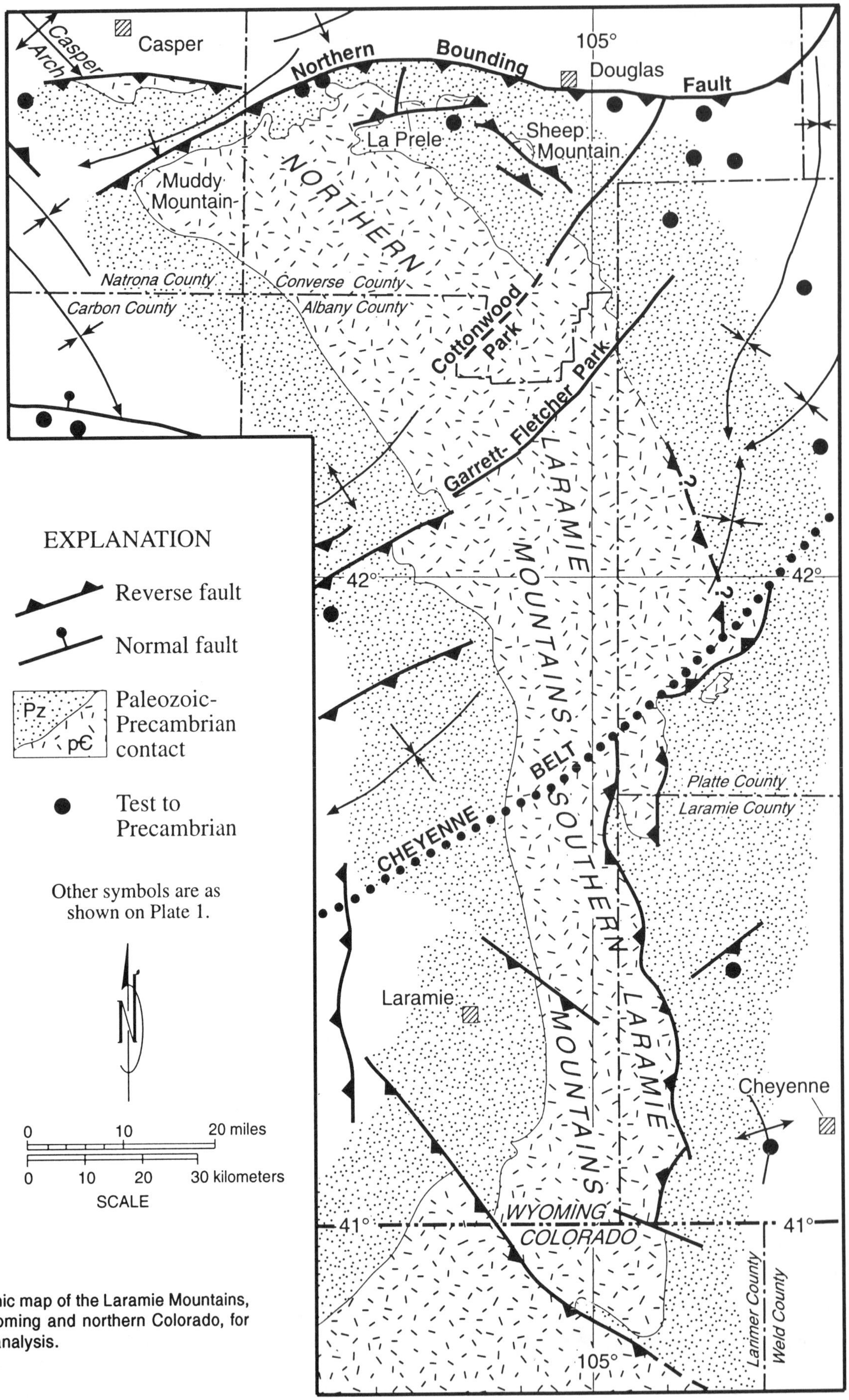

Figure 18. Tectonic map of the Laramie Mountains, southeastern Wyoming and northern Colorado, for use in structural analysis.

structure (Allmendinger and others, 1982). The reflection profiles revealed the details of crustal structure to a depth of 43.5 miles (70 km) including west dipping thrust faults and possible differences of crustal thickness on opposite sides of the Cheyenne belt.

Johnson and Smithson (1985) reprocessed the data from COCORP Line No. 3 and proposed a new interpretation of the fault situation bounding the east flank of the mountains. They concluded that what Brewer and others (1982) interpreted as sedimentary rocks extending 1.9 miles (3 km) beneath the thrust plate was the result of computer stacking of strong refraction energy rather than a sedimentary reflector. The interpretation provides a more rational fit of the surface geology (Hammond, 1949) with the seismic reflection data.

Speece (1992) interpreted a seismic reflection profile recorded near Horse Creek Siding on the Burlington Northern Railroad. His positioning of the thrust fault at the mountain front does not fit well with known surface geology.

Chronologic evolution of the Laramie Mountains

General stratigraphic framework

The present Laramie Mountains lay within an epeiric seaway during Cretaceous time (Weimer, 1959; Gill and Cobban, 1973; Lillegraven and Ostresh, 1988). The last major marine regression occurred during Maastrichtian time resulting in the deposition of the Fox Hill Sandstone, now found on both flanks of the Laramie Mountains (**Figure 5**).

Northeast of Rock River, Wyoming in T20N, R75W, the Fox Hills Sandstone is overlain unconformably by the lower part of the Medicine Bow Formation (Gill and others, 1970) of Lancian age. The Medicine Bow Formation thickens to the west into the Hanna Basin (Bowen, 1918) attaining a thickness of at least 4000 feet (1219 m). Fox (1971) found a few foraminifera in the Medicine Bow Formation near Rock River, Wyoming, indicating local marine conditions during deposition of this formation. The Medicine Bow Formation is overlain unconformably by the Paleocene Hanna Formation along the western flank of the Laramie Basin (Blackstone, 1983).

Local stratigraphic relationships

East of the Laramie Mountains along the valleys of Crow Creek and Horse Creek, the Fox Hills Sandstone crops out and dips at low angles to the east (Darton and others, 1910; Gray, 1947; Brady, 1949) (**Figure 3**). The strata overlying the Fox Hills Sandstone in the Denver-Cheyenne Basin are termed "Laramie Formation" in well records but no strata of this age appear on the map of the Laramie-Sherman Quadrangle (Darton and others, 1910). There is little hard evidence as to the actual age of the so-called Laramie Formation in the Denver-Cheyenne Basin (Weimer, 1959).

Measured sections of rocks on both flanks of the Laramie Mountains appear to contain rocks of Lancian age, indicating that the mountains did not begin to rise until at least late Lancian time (**Figure 5**).

On the west side of the Laramie Basin near Cooper Cove oil field (T18N, R77W), rocks of Paleocene age (Clarkforkian land mammal age), designated the Hanna Formation, unconformably overlie the late Cretaceous Lewis Shale (Blackstone, 1983). This relationship indicates that the northern Medicine Bow Mountains were elevated in early to middle Paleocene time and supplied some coarse clastic debris found in the basal Hanna Formation. Probably the Laramie Mountains were also elevated at this time. No rocks of Paleocene age are known on the eastern flank of the Laramie Mountains, suggesting that drainage from the western flank of the mountains was deflected into the present site of the Powder River Basin (Lillegraven and Ostresh, 1988).

Stratigraphy in northern Colorado

The northern Colorado Front Range between the Denver and North Park Basins was elevated at about the same time as the Medicine Bow Mountains (Hail and Kinney, 1959). The Paleocene Coalmont Formation in North Park Basin unconformably overlies the lower sandy phase of the Cretaceous Pierre Shale with no intervening Mesaverde Formation, Lewis Shale, Fox Hills Sandstone, or Medicine Bow Formation such as are found in the southern Laramie Basin (Blackstone, 1983).

These units were removed by erosion prior to the deposition of the Coalmont Formation. Since the Laramie Mountains are the northern extension of the Colorado Front Range it is likely they had a similar erosional history.

Conglomerates of uncertain age

Coarse clastic, relatively unconsolidated deposits crop out on the west flank of the Laramie Mountains northwest of State Highway 34 and northwest of Morton Pass (**Plate 1**). Boulders up to 6 feet (1.8 m) in diameter consist of anorthosite, gabbro, and syenitic rock types and occasional Paleozoic sedimentary rocks. Similar deposits cap the ridge west of Long Lake (Long Lake 7 1/2-minute Quadrangle) and contain numerous large blocks of Casper Sandstone. These deposits were shown on the geologic map of Wyoming (Love and others, 1955) as Paleocene Hanna Formation and on the 1985 version of the state map as Eocene Wagonbed Formation (Love and Christiansen, 1985). There is no hard (paleontological) evidence to establish the age of these deposits. The writer (Blackstone, 1975) considered them to probably be of Pleistocene (?) age.

Rocks of Eocene age

Rocks of Eocene age occupy the Cooper Lake synclinal basin (**Figure 1**). The rocks were named Wind River Formation by Nace (1936) on the basis of limited data. Later, Princhinello (1971) and Davidson (1987) confirmed the age of these rocks as Wasatchian (North American land mammal age). The strata include variegated mudstones, local conglomerates, limy concretionary intervals, and sandstones. Conglomerates cropping out in section 7, T19N, R74W, along the Laramie River contain pebbles of anorthosite and appear to be overlain by the coarse clastics described above.

The Wind River Formation in the Laramie Basin is folded into a broad, shallow syncline and unconformably overlies the Hanna Formation. No rocks of comparable character or age have been reported on the east flank of the Laramie Mountains. However, north of the mountains and north of the NBF the Eocene Wasatch Formation is widespread (Denson and Horn, 1975). The Laramie Mountains evidently were well defined in Eocene time, and a mature topography developed with deep valleys extending back into the core of the mountains. Evanoff (1990) demonstrated the backfilling of these valleys with fine-grained sediments of Chadronian North American land mammal age.

Oligocene age rocks

A major unconformity separates the Eocene-Oligocene White River Formation from all older rocks in southeastern Wyoming. Rocks assigned to the White River Formation crop out in North Park, Colorado (Evanoff, 1990), southern and northern Laramie Basin, and extensively east of the Laramie Mountains.

Present definition of the temporal boundary between epochs places the Chadronian time interval in the late Eocene. The basal upper part of the White River Formation is Chadronian in age and the upper part is Oligocene. The basal part of the White River contains mammalian fossils of Chadronian age (Evanoff, 1990; Moore, 1959). Assignment of the Wind River Formation in the Laramie Basin to the Eocene is based on collections of vertebrate fossils in the lower part of the formation (Nace, 1936; Princhinello, 1971; Davidson, 1987). No fossil collections have been from the uppermost part of the Wind River Formation that might correlate with the Chadronian age lower White River Formation east of the mountains. Until this discrepancy is resolved, the history of the Eocene will be in doubt.

The upper part of the White River Formation contains a high percentage of volcaniclastic debris derived from a remote source. Within this sequence are channel deposits of conglomerate containing rock types derived from the Precambrian core of the mountains (Stanley, 1971; Moore, 1959). Deposition continued through Oligocene and Miocene time until only the crest of the range stood above the aggraded surface. The Cenozoic history of southeastern Wyoming was discussed by Blackstone (1983) and need not be repeated here.

Tectonic overview

The bulk of this paper is a description of the structural elements of the Laramie Mountains with emphasis on the bounding (marginal) features. A compilation of structural elements at a scale suitable for synthesis is shown on **Figure 18**. Rocks of Precambrian age in the core of the mountains are highly fractured but the fracturing is at a scale too small to be shown on the generalized geologic map, **Plate 1**. The reader is referred to Bekkar (1973) for a map and discussion of the numerous linear fractures. Individual elements will be discussed from north to south. In the northwestern corner of the map area (**Plate 1**) the Casper Arch trends N45°W, essentially the same orientation as the core of the northern Laramie Mountains and associated en echelon folds. The northern limit of the mountain mass is a major reverse fault trending about N80°W across the map area herein named the Northern Bounding Fault (NBF). Casper Mountain anticline is a subparallel feature.

The northern Laramie Mountains are segmented by a series of northeast-striking shear zones and minor folds. The northernmost of these shear zones are the Cottonwood Park and Garrett-Fletcher Park shears (Chamberlin and others, 1993; Snyder, 1993). The Cheyenne belt and related features are the southernmost of the shear zones. The northeast orientation of the structures continues into the Hartville uplift in the northeastern corner of the map area (Denson and Botinelly, 1949).

The southern Laramie Mountains south of the Cheyenne belt (and Wheatland-Whalen fault zone) trend essentially north-south and continue 245 miles (394 km) into Colorado as the Colorado Front Range (Warner, 1978). The southern Laramie Mountain unit is bounded on the east by a series of west dipping reverse faults.

The southern Laramie Mountains are segmented by faults striking N50°W such as the Boulder Ridge fault (Beckwith, 1938), Granite Canyon area, and "The Spur" northeast of Laramie, Wyoming (Lundy, 1978). These faults dip steeply to the southwest and have fairly large displacements.

The overall mountain mass as shown on **Figure 18** is a major, broadly arched, asymmetric fold. The axis of regional shortening is oriented in general from southwest to northeast. Tectonic transport was from west to east or northeast. The same situation exists in the northern part of the Medicine Bow Mountains north of the Cheyenne belt and at Sheep Mountain, in Albany County south of the belt.

Directions of tectonic transport

Two anomalous directions of tectonic transport exist within the broad framework of the uplift. The first is the direction of movement on the NBF fault which is from south to north, crowding the core of the uplift northward toward the flank and axis of the Powder River Basin. Similar movement takes place on the Casper Mountain fault.

The second anomalous direction of tectonic transport is found in Platte County, Wyoming, approximately parallel to the Cheyenne belt. The Cheyenne belt was described earlier in this paper as a major suture of Precambrian age, possibly reactivated in Laramide time. The Mule Creek thrust fault parallels the trend of the Cheyenne belt and movement is to the southeast so that Precambrian basement is thrust over strata as young as the Cretaceous Mowry Shale. The northeast extension of this fault appears to bound the southeastern flank of the Hartville uplift.

A further anomalous situation exists in the Sheep Mountain-Deadhead area where thrusting is from west to east. That sense of motion is also present in the Dwyer Junction area several miles to the northwest. There is no hard evidence for strike slip (such as offset fold axes, etc.) movement along the Cheyenne belt. On the west flank of the Laramie Mountains, there is no offset of the contact between the Precambrian basement and the overlying Paleozoic sedimentary rocks.

Conclusions

1. The Laramie Mountains area major foreland uplift formed during the Laramide tectonic episode.
2. The geometric form of the uplift in cross section is that of long wavelength and low amplitude.
3. The uplift is "compartmentalized" along shear zones of Precambrian age, probably reactivated during Laramide deformation.

4. The rocks of Precambrian age in the mountain core are extensively fractured allowing for differential movement on many planes.

5. Variation in direction of tectonic transport probably reflects differences in response to a fairly uniform contractional stress field along older planes of failure.

6. The northern segment of the Laramie Mountains may have rotated in a clockwise direction.

7. The Northern Bounding Fault (NBF) is a major regional structural element separating the Laramie Mountains from the Powder River Basin.

8. Post-Miocene extensional faulting is superposed on all older structural features.

References

Ahlbrandt, T.S., and Groen, Wilson, 1987, The Goshen Hole uplift: a brief review of its geologic history and exploration potential: The Mountain Geologist, v. 24, p. 33-44.

Allmendinger, R.W., Brewer, J.A., Kaufmann, Steven, Oliver, J.E., and Houston, R.S., 1982, COCORP profiling across the Rocky Mountain Front in southern Wyoming-Part 2, Precambrian basement structure and its influence on Laramide deformation: Geological Society of America Bulletin, v. 93, p. 1253-1263.

Barlow, J.A., Jr., 1950, Geology of the La Prele-Box Elder Creek area, Converse County, Wyoming: M.A. thesis, University of Wyoming, Laramie, 49 p.

Beckwith, R.H., 1938, Structure of the southwest margin of the Laramie Basin, Wyoming: Geological Society of America Bulletin, v. 49, p. 1515-1544.

Bekkar, Hamed, 1973, Laramide structural elements and relationship to Precambrian basement, southeastern Wyoming: M.S. thesis, University of Wyoming, Laramie, 70 p.

Berryman, R.J., 1942, Geology of the Deer Creek-Little Deer Creek area, Converse County, Wyoming: M.A. thesis, University of Wyoming, Laramie, 51 p.

Biggs, Paul, and Espach, R.H., 1960, Petroleum and natural gas fields in Wyoming: U.S. Bureau of Mines Bulletin 582, 538 p.

Blackstone, D.L., Jr., 1975, Late Cretaceous and Cenozoic history of the Laramie Basin region: Geological Society of America Bulletin, v. 22, p. 249-278.

______, 1983, Laramide compressional tectonics, southeastern Wyoming: University of Wyoming Contributions to Geology, v. 22, p. 1-37.

______, 1988, Thrust faulting: southern margin of the Powder River Basin, Wyoming: Wyoming Geological Association 39th Annual Field Conference Guidebook, p. 35-44.

______, 1993a, Overview of the Hanna, Carbon, and Cooper Lake Basins, southeastern Wyoming: Wyoming State Geological Survey Report of Investigations 49, 20 p.

______, 1993b, Precambrian basement map of Wyoming: Outcrop and configuration: Wyoming State Geological Survey Map Series 43, scale 1:1,000,000.

______, 1993c, The Wind River Range: an overview: Wyoming Geological Association Jubilee Anniversary Field Conference Guidebook, p. 121-141.

Blackwelder, Elliot, 1909, Cenozoic history of the Laramie region, Wyoming: Journal of Geology, v. 17, p. 429-141.

Boos, C.M., and Boos, M.F., 1957, Tectonics of eastern flank and foothills of the Front Range, Colorado: American Association of Petroleum Geologists Bulletin, v. 41, p. 2603-2076.

Bowen, C.F., 1918, Stratigraphy of the Hanna Basin, Wyoming: U.S. Geological Survey Professional Paper 108-L, p. 227-235.

Brady, R.T., 1949, Geology of the east flank of the Laramie Range in the vicinity of Federal and Hecla, Laramie County, Wyoming: M.A. thesis, University of Wyoming, Laramie, 41 p.

Bretz, J.H., and Horberg, Leland, 1952, A high-level boulder deposit east of the Laramie Range, Wyoming: Journal of Geology, v. 60, p. 480-488.

Brewer, J.A., Allmendinger, R.W., Brown, L.D., Oliver, J.E., and Kaufman, Steven, 1982, COCORP profiling across the Rocky Mountain Front in southern Wyoming-Part I, Laramide structure: Geological Society of America Bulletin, v. 93, p. 1242-1252.

Chamberlain, R., Patel, S.C., Frost, B.R., and Snyder, George, 1993, Thick skinned deformation of the Archean Wyoming Province during Proterozoic arc-continent collision: Geology, v. 21, p. 995-998.

Condie, K.C., 1969, Petrology and geochemistry of the Laramie batholith and related metamorphic rocks of Precambrian age, eastern Wyoming: Geological Society of America Bulletin, v. 80, p. 57-82.

Copland, J.R., 1984, Laramide structural deformation at the interface between the Laramie Range and the Denver-Julesburg Basin, southeastern Wyoming: M.S. thesis, University of Wyoming, Laramie, 48 p.

Curry, W.H., and Curry, W.H., Jr., 1972, South Glenrock oil field Wyoming: Prediscovery thinking and post discovery description: American Association of Petroleum Geologists Memoir 16, p. 415-427.

Darton, N.H., Blackwelder, Elliot, and Siebenthal, C.E., 1910, Laramie-Sherman Folio, Wyoming: U.S. Geological Survey Geologic Atlas Folio 173, scale 1:125,000.

Davidson, J.R., 1987, Geology and mammalian paleontology of the Wind River Formation, Laramie Basin, southeastern Wyoming: M.S. thesis, University of Wyoming, Laramie, 101 p.

Denson, N.M., and Botinelly, Theodore, 1949, Tectonic map of the Hartville uplift and adjacent areas: Wyoming Geological Association 4th Annual Field Conference Guidebook, p. 41.

Denson, N.M., and Bergendahl, M.H., 1961, Middle and upper Tertiary rocks of southeastern Wyoming and adjoining areas: U.S. Geological Survey Professional Paper 424-C, p. 168-172.

Denson, N.M., and Horn, G.H., 1975, Geologic and structure map of the southern part of the Powder River Basin, Converse, Niobrara, and Natrona Counties, Wyoming: U.S. Geological Survey Miscellaneous Investigations Series Map I-877, scale 1:125,000.

Drouillard, E.K., 1963, Tectonics of the southeast flank of the Hartville uplift, Wyoming: Rocky Mountain Association of Geologists 14th Annual Field Conference Guidebook, p. 176-178.

Drwenski, V.R., 1952, Geology of the Boxelder-Mormon Canyon area, Converse County, Wyoming: M.A. thesis, University of Wyoming, Laramie, 67 p.

Evanoff, Emmet, 1990, Early Oligocene paleovalleys in southern and central Wyoming: Evidence of high local relief on the late Eocene unconformity: Geology, v. 18, p. 443-446.

Fox, J.E., 1971, Invertebrate fossils and environments of the Fox Hills and Medicine Bow Formations (Late Cretaceous) in south-central Wyoming: Ph.D. dissertation, University of Wyoming, Laramie, 161 p.

Gable, D.J., Buford, A.E., and Corbett, R.G., 1988, The Precambrian geology of Casper Mountain, Natrona County, Wyoming: U.S. Geological Survey Professional Paper 1460, 50 p.

Gill, J.R., Merewether, E.A., and Cobban, W.A., 1970, Stratigraphy and nomenclature of some Upper Cretaceous and lower Tertiary rocks in south-central Wyoming: U.S. Geological Survey Professional Paper 667, 53 p.

Gill, J.R., and Cobban, W.A., 1973, Stratigraphy and geologic history of the Montana Group and equivalent rocks, Montana, Wyoming, and North and South Dakota: U.S. Geological Survey Professional Paper 776, 37 p.

Graff, P.J., Sears, J.W., Holden, G.S., and Hausel, W.D., 1982, Geology of the Elmer's Rock greenstone belt, Laramie Range, Wyoming: Wyoming State Geological Survey Report of Investigations 14, 25 p.

Gray, L.O., Jr., 1947, Geology of the east flank of the Laramie Range in the vicinity of Horse Creek, Laramie County, Wyoming: M.A. thesis, University of Wyoming, Laramie, 25 p.

Hammond, C.R., 1949, Geology of the east flank of the Laramie Range in the vicinity of Iron Mountain, Wyoming: M.A. thesis, University of Wyoming, Laramie, 63 p.

Hail, W.J., Jr., and Kinney, D.U., 1959, Upper Cretaceous rocks in North Park, Jackson County, Colorado, *in* Haun, J.D., and Weimer, R.J., editors, Symposium on Cretaceous rocks of Colorado and adjacent regions: Rocky Mountain Association of Geologists, Denver, Colorado, p. 105-110.

Haun, J.D., 1949, Geology of a portion of the east flank of the Laramie Range, Wyoming: M.A. thesis, University of Wyoming, Laramie, 40 p.

Hills, F.A., and Armstrong, R.L., 1974, Geochronology of Precambrian rocks in the Laramie Range and implications for tectonic framework of Precambrian in southern Wyoming; Precambrian Research, v. 1, p. 213-225.

Hodge, Dennis, Owen, L.B., and Smithson, S.B., 1973, Gravity interpretation of the Laramie anorthosite complex, Wyoming: Geological Society of America Bulletin, v. 84, p. 1451-1464.

Houston, R.S., and others, 1968, A regional study of rocks of Precambrian age in that part of the Medicine Bow Mountains lying in southeastern Wyoming *with a chapter on* the relationship between Precambrian and Laramide structure: Wyoming State Geological Survey Memoir 1, 165 p.

Hunter, Z.M., 1955, Geology-map of the foothills of the Front Range of northern Colorado: Rocky Mountain Association of Geologists, map.

Iltis, S.T., 1983, Processing and interpretation of seismic reflection data from the Precambrian of the central Laramie Range, Albany County, Wyoming: M.S. thesis, University of Wyoming, Laramie, 94 p.

Jenkins, C.E., and Rea, B.D., 1978, A field guide to the Casper Mountain area: The Wyoming Field Science Foundation: Casper, Wyoming, 83 p.

Jenkins, P.T., 1938, Geology of a portion of the east side of the Laramie Range, Albany, Laramie, and Platte Counties, Wyoming: M.A. thesis, University of Wyoming, Laramie, 42 p.

Johnson, R.C., and Hills, F.A., 1976, Precambrian geochronology and geology of the Box Elder Canyon area, northern Laramie Range, Wyoming: Geological Society of America Bulletin, v. 87, p. 809-817.

Johnson, Roy, and Smithson, Scott, 1985, Thrust faulting in the Laramie Mountains, Wyoming, from reanalysis of COCORP data: Geology, v. 13, p. 534-546.

Karlstrom, Karl, and Houston, R.S., 1984, The Cheyenne Belt: Analysis of a Proterozoic suture in southern Wyoming: Precambrian Research, v. 25, p. 415-446.

Knight, W.H., 1944, The geology of the western part of the Flat Top anticline, Carbon County, Wyoming: M.A. thesis, University of Wyoming, Laramie, 30 p.

Langstaff, G.D., 1988, Investigation of Archean metavolcanic and metasedimentary rocks of Seller's Mountain, west central Laramie Mountains, Wyoming: Ph.D. dissertation, University of Wyoming, Laramie, 384 p.

Lillegraven, J.A., 1993, Correlation of Paleogene strata across Wyoming: A user's guide, *in* Snoke, A.W., Steidtmann, J.R., and Roberts, S.M., editors, Geology of Wyoming: Wyoming State Geological Survey Memoir 5, v. 2, p. 414-480.

Lillegraven, J.A., and Ostresh, L.M., Jr., 1988, Evolution of Wyoming's early Cenozoic topography and drainage patterns: National Geographic Research, v. 4, p. 303-327.

Love, J.D., and Christiansen, A.C., 1985, Geologic map of Wyoming: U.S. Geological Survey, scale 1:500,000.

Love, J.D., Denson, N.M., and Botinelly, Theodore, 1949, Geology of the Glendo area, Wyoming: U.S. Geological Survey Oil and Gas Investigations Map 92, scale 1:48,000.

Love, J.D., Weitz, J.L., and Hose, R.K., 1955, Geologic map of Wyoming: U.S. Geological Survey, scale 1:500,000.

Loeffler, R.J., 1939, The stratigraphy of a portion of the east side of the Laramie Range, Laramie County, Wyoming: M.A. thesis, University of Wyoming, Laramie, 45 p.

Lovering, T.S., 1929, Geologic history of the Front Range: Colorado Scientific Society Proceedings, v. 12, p. 59-111.

Lundy, D., 1978, Hydrology and geochemistry of the Casper acquifer in the vicinity of Laramie, Wyoming: Ph.D. dissertation, University of Wyoming, Laramie.

Lynn, J.R., 1947, Geology of the north margin of the Reshaw Hills, Platte County, Wyoming: M.A. thesis, University of Wyoming, Laramie, 53 p.

McCurdy, H.R., 1941, Pennsylvanian and Permian stratigraphy along the eastern flank of the Laramie Range, southeastern Wyoming: M.A. thesis, University of Wyoming, Laramie, 45 p.

McGookey, D.P., 1952, Pennsylvanian and early Permian stratigraphy of the southeast portion of the east flank of the Laramie Range, Wyoming and Colorado: M.S. thesis, University of Wyoming, Laramie, 45 p.

McGrew, L.W., 1953, The geology of the Grayrocks area, Platte and Goshen Counties, Wyoming: M.S. thesis, University of Wyoming, Laramie, 94 p.

______, 1963, Geology of the Fort Laramie area, Platte and Goshen Counties, Wyoming: U.S. Geological Survey Bulletin 1141-F, 39 p.

______, 1967a, Geologic map of the Natwick Southwest Quadrangle, Platte County, Wyoming: U.S. Geological Survey Geologic Quadrangle Map GQ-623, scale 1:24,000.

______, 1967b, Geologic map of the Richeau Hills Quadrangle, Platte County, Wyoming: U.S. Geological Survey Geologic Quadrangle Map GQ-625, scale 1:24,000.

McMinn, P.M., 1963, Geology of the Borie oil field, Laramie County, Wyoming: Rocky Mountain Association of Geologists 14th Annual Field Conference Guidebook, p. 234-235.

Minick, J.N., 1951, Tertiary stratigraphy of southeastern Wyoming and northeastern Colorado: M.S. thesis, University of Wyoming, Laramie, 53 p.

Moore, F.E., 1959, The geomorphic evolution of the east flank of the Laramie Range, Colorado and Wyoming: Ph.D. dissertation, University of Wyoming, Laramie, 123 p.

Mort, F.P., 1939, The geology of an area between La Bonte and La Prele Creeks, Converse County, Wyoming: M.S. thesis, University of Wyoming, Laramie, 16 p.

Mueller, R.E., 1982, The Cheyenne Belt, southeastern Wyoming—Part 1: Descriptive geology and petrography: Ph.D. dissertation, University of Wyoming, Laramie, 98 p.

Newhouse, W.H., and Hagner, A.F., 1957, Geologic map of anorthosie areas, southern part of Laramie Range, Wyoming: U.S. Geological Survey Miscellaneous Field Studies Map 119, scale 1:63,360.

Nace, R.L., 1936, Summary of Late Cretaceous and early Tertiary stratigraphy of Wyoming: Wyoming State Geological Survey Bulletin 26, 271 p.

Peterman, Z.E., and Hildreth, R.A., 1978, Reconnaissance geology and geochronology of the Precambrian of the Granite Mountains, Wyoming: U.S. Geological Survey Professional Paper 1055, 22 p.

Princhinello, K.A., 1971, Earliest Eocene mammalian fossils from Laramie Basin of southeastern Wyoming: M.S. thesis, University of Wyoming, Laramie, 39 p.

Ransome, F.L., 1915, Tertiary orogeny of the North American Cordillera and its problems, *in* Problems of American Geology, New Haven, Connecticut, p. 287-376.

Rapp, J.R., and Babcock, H.M., 1953, Reconnaissance of the geology and groundwater resources of the Glendo-Wendover area, Platte County, Wyoming: U.S. Geological Survey Circular 163, 34 p.

Sando, W.J., and Sandberg, C.A., 1987, New interpretation of Paleozoic stratigraphy and history of the northern Laramie Range and vicinity, southeast Wyoming: U.S. Geological Survey Professional Paper 1450, 38 p.

Schwarberg, T.M., Jr., 1959, The geology of Muddy Mountain, southeastern Natrona County, Wyoming: M.S. thesis, University of Wyoming, Laramie, 103 p.

Sears, W.A., Jr., 1949, Geology of the Deer Creek-Smith Creek area, Converse County Wyoming: M.S. thesis, University of Wyoming, Laramie, 56 p.

Segerstrom, Kenneth, and Weisner, R.C., 1977, Mineral resources of the Laramie Peak study area, Albany and Converse Counties, Wyoming: U.S. Geological Survey Bulletin 1397-B, 35 p.

Shorey, J.A., and Howell, D.H., 1963, Gypsum deposits of Northern Colorado Front Range: Rocky Mountain Association of Geologists 14th Annual Field Conference Guidebook, p. 65-66.

Smith, W.S.T., 1903, Hartville Folio, Wyoming: U.S. Geological Survey Geologic Atlas Folio 91, scale 1:125,000.

Smithson, S.B., Brewer, J.A., Kaufman, Steven, Oliver, J.A., and Hurich, C.A., 1979, Structure of the Laramide Wind River uplift, Wyoming, from COCORP deep reflection data and from gravity data: Journal of Geophysical Research, v. 84, p. 5955-5972.

Snyder, G.L. 1993, Geologic map of the Esterbrook-Braae area, Albany, Converse, and Platte Counties, Wyoming: U.S. Geological Survey Miscellaneous Investigations Series Map I-2232, scale 1:24,000.

Speece, M.A., 1992, Geophysical studies in Precambrian regions: the Laramie Mountains, Wyoming, and the Godthabsfjord area, southern west Greenland: Ph.D. dissertation, University of Wyoming, Laramie, 110 p.

Speece, M.A., Frost, B.R., and Smithson, S.B., 1994, Precambrian basement structure and Laramide deformation revealed by seismic reflection profiling in Laramie Mountains, Wyoming: Tectonics, v. 13, p. 354-366.

Spelman, A.R., 1959, Geology of the area between Bed Tick Creek and the west fork of La Bonte Creek, Converse County, Wyoming: M.S. thesis, University of Wyoming, Laramie, 81 p.

Spencer, A.C., 1916, The Atlantic gold district and north Laramie Mountains, Fremont, Converse, and Albany Counties, Wyoming: U.S. Geological Survey Bulletin 616, 85 p.

Stanley, K.O., 1971, Tectonic implications of Tertiary sediment dispersal on the Great Plains east of the Laramie Range: Wyoming Geological Association 23rd Annual Field Conference Guidebook, p. 65-71.

Thompson, W.O., 1938, Structural map of Sand Creek and Box Elder Creek anticline, Larimer County, Colorado: Kansas Geological Society 12th Annual Field Conference Guidebook, p. 65-66.

Toots, H.A., 1965, Reconstruction of continental environments, the Oligocene of Wyoming: Ph.D. dissertation, University of Wyoming, Laramie, 175 p.

Tudor, M.S., 1952, Geology of the west-central flank of the Laramie Range, Albany County, Wyoming: M.S. thesis, University of Wyoming, Laramie, 51 p.

Warner, L.A., 1978, The Colorado Lineament: a middle Precambrian wrench fault system: Geological Society of America Bulletin, v. 89, p. 161-171.

Weimer, R.J., 1959, Upper Cretaceous stratigraphy, Colorado: Rocky Mountain Association of Geologists 11th Annual Field Conference Guidebook, p. 9-16.

Wyoming Geological Association, 1984, Southeast Wyoming Oil and Gas Fields Symposium—Laramie, Hanna, and D-J Basins, 133 p.